XK
3.4, 3.8 & 4.2 LITRE ENGINES

Other great books from Veloce –

Speedpro Series
4-cylinder Engine – How To Blueprint & Build A Short Block For High Performance (Hammill)
Alfa Romeo DOHC High-performance Manual (Kartalamakis)
Alfa Romeo V6 Engine High-performance Manual (Kartalamakis)
BMC 998cc A-series Engine – How To Power Tune (Hammill)
1275cc A-series High-performance Manual (Hammill)
Camshafts – How To Choose & Time Them For Maximum Power (Hammill)
Competition Car Datalogging Manual, The (Templeman)
Cylinder Heads – How To Build, Modify & Power Tune Updated & Revised Edition (Burgess & Gollan)
Distributor-type Ignition Systems – How To Build & Power Tune New 3rd Edition (Hammill)
Fast Road Car – How To Plan And Build Revised & Updated Colour New Edition (Stapleton)
Ford SOHC 'Pinto' & Sierra Cosworth DOHC Engines – How To Power Tune Updated & Enlarged Edition (Hammill)
Ford V8 – How To Power Tune Small Block Engines (Hammill)
Harley-Davidson Evolution Engines – How To Build & Power Tune (Hammill)
Holley Carburetors – How To Build & Power Tune Revised & Updated Edition (Hammill)
Honda Civic Type R, High-Performance Manual ((Cowland & Clifford)
Jaguar XK Engines – How To Power Tune Revised & Updated Colour Edition (Hammill)
MG Midget & Austin-Healey Sprite – How To Power Tune New 3rd Edition (Stapleton)
MGB 4-cylinder Engine – How To Power Tune (Burgess)
MGB V8 Power – How To Give Your, Third Colour Edition (Williams)
MGB, MGC & MGB V8 – How To Improve New 2nd Edition (Williams)
Mini Engines – How To Power Tune On A Small Budget Colour Edition (Hammill)
Motorcycle-engined Racing Car – How To Build (Pashley)
Motorsport – Getting Started in (Collins)
Nissan GT-R High-performance Manual, The(Gorodji)
Nitrous Oxide High-performance Manual, The (Langfield)
Rover V8 Engines – How To Power Tune (Hammill)
Sportscar & Kitcar Suspension & Brakes – How To Build & Modify Revised 3rd Edition (Hammill)
SU Carburettor High-performance Manual (Hammill)
Successful Low-Cost Rally Car, How to Build a (Young)
Suzuki 4x4 – How To Modify For Serious Off-road Action (Richardson)
Tiger Avon Sportscar – How To Build Your Own Updated & Revised 2nd Edition (Dudley)
TR2, 3 & TR4 – How To Improve (Williams)
TR5, 250 & TR6 – How To Improve (Williams)
TR7 & TR8 – How To Improve (Williams)
V8 Engine – How To Build A Short Block For High Performance (Hammill)
Volkswagen Beetle Suspension, Brakes & Chassis – How To Modify For High Performance (Hale)
Volkswagen Bus Suspension, Brakes & Chassis – How To Modify For High Performance (Hale)
Weber DCOE, & Dellorto DHLA Carburetors – How To Build & Power Tune 3rd Edition (Hammill)

Those Were The Days ... Series
Alpine Trials & Rallies 1910-1973 (Pfundner)
American Trucks of the 1950s (Mort)
Anglo-American Cars From the 1930s to the 1970s (Mort)
Austerity Motoring (Bobbitt)
Austins, The last real (Peck)
Brighton National Speed Trials (Gardiner)
British Lorries Of The 1950s (Bobbitt)
British Lorries of the 1960s (Bonnitt)
British Touring Car Championship, The (Collins)
British Police Cars (Walker)
British Woodies (Peck)
Café Racer Phenomenon, The (Walker)
Dune Buggy Phenomenon (Hale)
Dune Buggy Phenomenon Volume 2 (Hale)
Hot Rod & Stock Car Racing in Britain In The 1980s (Neil)
Last Real Austins, The, 1946-1959 (Peck)
MG's Abingdon Factory (Moylan)
Motor Racing At Brands Hatch In The Seventies (Parker)
Motor Racing At Brands Hatch In The Eighties (Parker)
Motor Racing At Crystal Palace (Collins)
Motor Racing At Goodwood In The Sixties (Gardiner)
Motor Racing At Nassau In The 1950s & 1960s (O'Neil)
Motor Racing At Oulton Park In The 1960s (McFadyen)
Motor Racing At Oulton Park In The 1970s (McFadyen)
Superprix (Collins)
Three Wheelers (Bobbitt)

Enthusiast's Restoration Manual Series
Citroën 2CV, How To Restore (Porter)
Classic Car Bodywork, How To Restore (Thaddeus)
Classic British Car Electrical Systems (Astley)
Classic Car Electrics (Thaddeus)
Classic Cars, How To Paint (Thaddeus)
Reliant Regal, How To Restore (Payne)
Triumph TR2, 3, 3A, 4 & 4A, How To Restore (Williams)
Triumph TR5/250 & 6, How To Restore (Williams)
Triumph TR7/8, How To Restore (Williams)
Volkswagen Beetle, How To Restore (Tyler)
VW Bay Window Bus (Paxton)
Yamaha FS1-E, How To Restore (Watts)

Essential Buyer's Guide Series
Alfa GT (Booker)
Alfa Romeo Spider Giulia (Booker & Talbott)
BMW GS (Henshaw)
BSA Bantam (Henshaw)
BSA Twins (Henshaw)
Citroën 2CV (Paxton)
Citroën ID & DS (Heilig)
Fiat 500 & 600 (Bobbitt)
Ford Capri (Paxton)
Jaguar E-type 3.8 & 4.2-litre (Crespin)
Jaguar E-type V12 5.3-litre (Crespin)
Jaguar XJ 1995-2003 (Crespin)
Jaguar/Daimler XJ6, XJ12 & Sovereign (Crespin)
Jaguar/Daimler XJ40 (Crespin)
Jaguar XJ-S (Crespin)
MGB & MGB GT (Williams)
Mercedes-Benz 280SL-560DSL Roadsters (Bass)
Mercedes-Benz 'Pagoda' 230SL, 250SL & 280SL Roadsters & Coupés (Bass)
Mini (Paxton)
Morris Minor & 1000 (Newell)
Porsche 928 (Hemmings)
Rolls-Royce Silver Shadow & Bentley T-Series (Bobbitt)
Subaru Impreza (Hobbs)
Triumph Bonneville (Henshaw)
Triumph Stag (Mort & Fox)
Triumph TR6 (Williams)
VW Beetle (Cservenka & Copping)
VW Bus (Cservenka & Copping)
VW Golf GTI (Cservenka & Copping)

Auto-Graphics Series
Fiat-based Abarths (Sparrow)
Jaguar MKI & II Saloons (Sparrow)
Lambretta Li Series Scooters (Sparrow)

Rally Giants Series
Audi Quattro (Robson)
Austin Healey 100-6 & 3000 (Robson)
Fiat 131 Abarth (Robson)
Ford Escort MkI (Robson)
Ford Escort RS Cosworth & World Rally Car (Robson)
Ford Escort RS1800 (Robson)
Lancia Stratos (Robson)
Mini Cooper/Mini Cooper S (Robson)
Peugeot 205 T16 (Robson)
Subaru Impreza (Robson)
Toyota Celica GT4 (Robson)

WSC Giants
Ferrari 312P & 312PB (Collins & McDonough)

Battle Cry! Original Military Uniforms of the World
Soviet General & field rank officers uniforms: 1955 to 1991

General
1½-litre GP Racing 1961-1965 (Whitelock)
AC Two-litre Saloons & Buckland Sportscars (Archibald)
Alfa Romeo Giulia Coupé GT & GTA (Tipler)
Alfa Romeo Montreal – The dream car that came true (Taylor)
Alfa Romeo Montreal – The Essential Companion (Taylor)
Alfa Tipo 33 (McDonough & Collins)
Alpine & Renault – The Development Of The Revolutionary Turbo F1 Car 1968 to 1979 (Smith)
Anatomy Of The Works Minis (Moylan)
André Lefebvre, and the cars he created at Voisin and Citroën (Beck)
Armstrong-Siddeley (Smith)
Autodrome (Collins & Ireland)
Automotive A-Z, Lane's Dictionary Of Automotive Terms (Lane)
Automotive Mascots (Kay & Springate)
Bahamas Speed Weeks, The (O'Neil)
Bentley Continental, Corniche And Azure (Bennett)
Bentley MkVI, Rolls-Royce Silver Wraith, Dawn & Cloud/Bentley R & S-Series (Nutland)
BMC Competitions Department Secrets (Turner, Chambers & Browning)
BMW 5-Series (Cranswick)
BMW Z-Cars (Taylor)
BMW Boxer Twins 1970-1995 Bible, The (Falloon)
Britains Farm Model Balers & Combines 1967-2007, Pocket Guide to (Pullen)
Britains Farm Model & Toy Tractors 1998-2008, Pocket Guide to (Pullen)
British 250cc Racing Motorcycles (Pereira)
British Cars, The Complete Catalogue Of, 1895-1975 (Culshaw & Horrobin)
BRM – A Mechanic's Tale (Salmon)
BRM V16 (Ludvigsen)
BSA Bantam Bible, The (Henshaw)
Bugatti Type 40 (Price)
Bugatti 46/50 Updated Edition (Price & Arbey)
Bugatti T44 & T49 (Price & Arbey)
Bugatti 57 2nd Edition (Price)
Caravans, The Illustrated History 1919-1959 (Jenkinson)
Caravans, The Illustrated History From 1960 (Jenkinson)
Carrera Panamericana, La (Tipler)
Chrysler 300 – America's Most Powerful Car 2nd Edition (Ackerson)
Chrysler PT Cruiser (Ackerson)
Citroën DS (Bobbitt)
Classic British Car Electrical Systems (Astley)
Cliff Allison – From The Fells To Ferrari (Gauld)
Cobra – The Real Thing! (Legate)
Concept Cars, How to illustrate and design (Dewey)
Cortina – Ford's Bestseller (Robson)
Coventry Climax Racing Engines (Hammill)
Daimler SP250 New Edition (Long)
Datsun Fairlady Roadster To 280ZX – The Z-Car Story (Long)
Diecast Toy Cars of the 1950s & 1960s (Ralston)
Dino – The V6 Ferrari (Long)
Dodge Challenger & Plymouth Barracuda (Grist)
Dodge Charger – Enduring Thunder (Ackerson)
Dodge Dynamite! (Grist)
Donington (Boddy)
Draw & Paint Cars – How To (Gardiner)
Drive On The Wild Side, A – 20 Extreme Driving Adventures From Around The World (Weaver)
Ducati 750 Bible, The (Falloon)
Ducati 860, 900 And Mille Bible, The (Falloon)
Dune Buggy, Building A – The Essential Manual (Shakespeare)
Dune Buggy Files (Hale)
Dune Buggy Handbook (Hale)
Edward Turner: The Man Behind The Motorcycles (Clew)
Fast Ladies – Female Racing Drivers 1888 to 1970 (Bouzanquet)
Fiat & Abarth 124 Spider & Coupé (Tipler)
Fiat & Abarth 500 & 600 2nd Edition (Bobbitt)
Fiats, Great Small (Ward)
Fine Art Of The Motorcycle Engine, The (Peirce)
Ford F100/F150 Pick-up 1948-1996 (Ackerson)
Ford F150 Pick-up 1997-2005 (Ackerson)
Ford GT – Then, And Now (Streather)
Ford GT40 (Legate)
Ford In Miniature (Olson)
Ford Model Y (Roberts)
Ford Thunderbird From 1954, The Book Of The (Long)
Formula 5000 Motor Racing, Back then ... and back now (Lawson)
Forza Minardi! (Vigar)
Funky Mopeds (Skelton)
Gentleman Jack (Gauld)
GM In Miniature (Olson)
GT – The World's Best GT Cars 1953-73 (Dawson)
Hillclimbing & Sprinting – The Essential Manual (Short & Wilkinson)
Honda NSX (Long)
Intermeccanica - The Story of the Prancing Bull (McCredie & Reisner)
Jaguar, The Rise Of (Price)
Jaguar XJ-S (Long)
Jeep CJ (Ackerson)
Jeep Wrangler (Ackerson)
John Chatham - 'Mr Big Healey' – The Official Biography (Burr)
Karmann-Ghia Coupé & Convertible (Bobbitt)
Lamborghini Miura Bible, The (Sackey)
Lambretta Bible, The (Davies)
Lancia 037 (Collins)
Lancia Delta HF Integrale (Blaettel & Wagner)
Land Rover, The Half-ton Military (Cook)
Laverda Twins & Triples Bible 1968-1986 (Falloon)
Lea-Francis Story, The (Price)
Lexus Story, The (Long)
little book of smart, the New Edition (Jackson)
Lola – The Illustrated History (1957-1977) (Starkey)
Lola – All The Sports Racing & Single-seater Racing Cars 1978-1997 (Starkey)
Lola T70 – The Racing History & Individual Chassis Record 4th Edition (Starkey)
Lotus 49 (Oliver)
Marketingmobiles, The Wonderful Wacky World Of (Hale)
Mazda MX-5/Miata 1.6 Enthusiast's Workshop Manual (Grainger & Shoemark)
Mazda MX-5/Miata 1.8 Enthusiast's Workshop Manual (Grainger & Shoemark)
Mazda MX-5 Miata: The Book Of The World's Favourite Sportscar (Long)
Mazda MX-5 Miata Roadster (Long)
Maximum Mini (Booij)
MGA (Price Williams)
MGB & MGB GT– Expert Guide (Auto-doc Series) (Williams)
MGB Electrical Systems Updated & Revised Edition (Astley)
Micro Caravans (Jenkinson)
Micro Trucks (Mort)
Microcars At Large! (Quellin)
Mini Cooper – The Real Thing! (Tipler)
Mitsubishi Lancer Evo, The Road Car & WRC Story (Long)
Montlhéry, The Story Of The Paris Autodrome (Boddy)
Morgan Maverick (Lawrence)
Morris Minor, 60 Years On The Road (Newell)
Moto Guzzi Sport & Le Mans Bible, The (Falloon)
Motor Movies – The Posters! (Veysey)
Motor Racing – Reflections Of A Lost Era (Carter)
Motorcycle Apprentice (Cakebread)
Motorcycle Road & Racing Chassis Designs (Noakes)
Motorhomes, The Illustrated History (Jenkinson)
Motorsport In colour, 1950s (Wainwright)
Nissan 300ZX & 350Z – The Z-Car Story (Long)
Nissan GT-R Supercar: Born to race (Gorodji)
Off-Road Giants! – Heroes of 1960s Motorcycle Sport (Westlake)
Pass The Theory And Practical Driving Tests (Gibson & Hoole)
Peking To Paris 2007 (Young)
Plastic Toy Cars Of The 1950s & 1960s (Ralston)
Pontiac Firebird (Cranswick)
Porsche Boxster (Long)
Porsche 356 (2nd Edition) (Long)
Porsche 908 (Födisch, Neßhöver, Roßbach, Schwarz & Roßbach)
Porsche 911 Carrera – The Last Of The Evolution (Corlett)
Porsche 911R, RS & RSR, 4th Edition (Starkey)
Porsche 911 – The Definitive History 1963-1971 (Long)
Porsche 911 – The Definitive History 1971-1977 (Long)
Porsche 911 – The Definitive History 1977-1987 (Long)
Porsche 911 – The Definitive History 1987-1997 (Long)
Porsche 911 – The Definitive History 1997-2004 (Long)
Porsche 911SC 'Super Carrera' – The Essential Companion (Streather)
Porsche 914 & 914-6: The Definitive History Of The Road & Competition Cars (Long)
Porsche 924 (Long)
Porsche 928 (Long)
Porsche 944 (Long)
Porsche 964, 993 & 996 Data Plate Code Breaker (Streather)
Porsche 993 'King Of Porsche' – The Essential Companion (Streather)
Porsche 996 'Supreme Porsche' – The Essential Companion (Streather)
Porsche Racing Cars – 1953 To 1975 (Long)
Porsche Racing Cars – 1976 To 2005 (Long)
Porsche – The Rally Story (Meredith)
Porsche: Three Generations Of Genius (Meredith)
RAC Rally Action! (Gardiner)
Rallye Sport Fords: The Inside Story (Moreton)
Redman, Jim – 6 Times World Motorcycle Champion: The Autobiography (Redman)
Rolls-Royce Silver Shadow/Bentley T Series Corniche & Camargue Revised & Enlarged Edition (Bobbitt)
Rolls-Royce Silver Spirit, Silver Spur & Bentley Mulsanne 2nd Edition (Bobbitt)
Russian Motor Vehicles (Kelly)
RX-7 – Mazda's Rotary Engine Sportscar (Updated & Revised New Edition) (Long)
Scooters & Microcars, The A-Z Of Popular (Dan)
Scooter Lifestyle (Grainger)
Singer Story: Cars, Commercial Vehicles, Bicycles & Motorcycle (Atkinson)
SM – Citroën's Maserati-engined Supercar (Long & Claverol)
Speedway – Motor Racing's Ghost Tracks (Collins & Ireland)
Subaru Impreza: The Road Car And WRC Story (Long)
Supercar, How To Build your own (Thompson)
Tales from the Toolbox (Oliver)
Taxi! The Story Of The 'London' Taxicab (Bobbitt)
Tinplate Toy Cars Of The 1950s & 1960s (Ralston)
Toleman Story, The (Hilton)
Toyota Celica & Supra, The Book Of Toyota's Sports Coupés (Long)
Toyota MR2 Coupés & Spyders (Long)
Triumph Bonneville!, Save the – The inside story of the Meriden workers' co-op (Rosamund)
Triumph Motorcycles & The Meriden Factory (Hancox)
Triumph Speed Twin & Thunderbird Bible (Woolridge)
Triumph Tiger Cub Bible (Estall)
Triumph Trophy Bible (Woolridge)
Triumph TR6 (Kimberley)
Unraced (Collins)
Velocette Motorcycles – MSS To Thruxton Updated & Revised (Burris)
Virgil Exner – Visioneer: The Official Biography Of Virgil M Exner Designer Extraordinaire (Grist)
Volkswagen Bus Book, The (Bobbitt)
Volkswagen Bus Or Van To Camper, How To Convert (Porter)
Volkswagens Of The World (Glen)
VW Beetle Cabriolet (Bobbitt)
VW Beetle – The Car Of The 20th Century (Copping)
VW Bus – 40 Years Of Splitties, Bays & Wedges (Copping)
VW Bus Book, The (Bobbitt)
VW Golf: Five Generations Of Fun (Copping & Cservenka)
VW – The Air-cooled Era (Copping)
VW T5 Camper Conversion Manual (Porter)
VW Campers (Copping)
Works Minis, The Last (Purves & Brenchley)
Works Rally Mechanic (Moylan)

www.veloce.co.uk

First published in 2001.Reprinted 2001 & 2005. This edition published April 2009 by Veloce Publishing Limited, 33 Trinity Street, Dorchester DT1 1TT, England. Fax 01305 268864/e-mail info@veloce.co.uk/web www.veloce.co.uk or www.velocebooks.com.
ISBN: 978-1-845840-05-1/UPC: 6-36847-04005-5
© Des Hammill and Veloce Publishing 2009. All rights reserved. With the exception of quoting brief passages for the purpose of review, no part of this publication may be recorded, reproduced or transmitted by any means, including photocopying, without the written permission of Veloce Publishing Ltd. Throughout this book logos, model names and designations, etc, have been used for the purposes of identification, illustration and decoration. Such names are the property of the trademark holder as this is not an official publication.
Readers with ideas for automotive books, or books on other transport or related hobby subjects, are invited to write to the editorial director of Veloce Publishing at the above address.
British Library Cataloguing in Publication Data – A catalogue record for this book is available from the British Library. Typesetting, design and page make-up all by Veloce Publishing Ltd on Apple Mac. Printed in India by Replika Press.

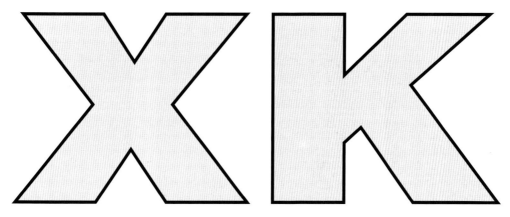

3.4, 3.8 & 4.2 LITRE ENGINES

Des Hammill

VELOCE PUBLISHING
THE PUBLISHER OF FINE AUTOMOTIVE BOOKS

Veloce SpeedPro books -

978-1-84584-142-3

978-1-84584-162-1

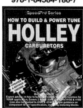
978-1-84584-186-7

978-1-84584-187-4

978-1-84584-207-9

978-1-84584-208-6

978-1-84584-224-6

978-1-845840-05-1

978-1-845840-06-8

978-1-845840-19-8

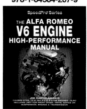
978-1-845840-21-1

978-1-845840-23-5

978-1-845840-45-7

978-1-845840-73-0

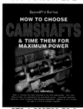
978-1-845841-23-2

978-1-845842-66-6

978-1-874105-70-1

978-1-901295-26-9

978-1-903706-14-5

978-1-903706-17-6

978-1-903706-59-6

978-1-903706-68-8

978-1-903706-70-1

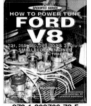
978-1-903706-72-5

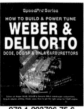
978-1-903706-75-6

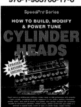
978-1-903706-76-3

978-1-903706-77-0

978-1-903706-78-7

978-1-903706-80-0

978-1-903706-92-3

978-1-903706-94-7

978-1-903706-99-2

978-1-904788-22-5

978-1-904788-78-2

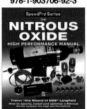
978-1-904788-84-3

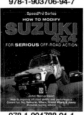
978-1-904788-91-1

978-1-904788-93-5

More on the way! ...

Contents

Thanks & introduction 7
Using this book & essential
  information 12

Chapter 1. Cylinder block 14
General checks & advice 14
Compression ratio (CR) 20
Petrol/gasoline for Jaguar XK
  engines 21
3.4-litre cylinder block - notes 23
3.8-litre cylinder block - notes 22
4.2-litre cylinder block - notes 22
Long head stud blocks 22
All short stud blocks 23
Freeze (Welch) plugs - strapping 24
Main bearing crush - checking 24
Main cap location dowels 27
Thrust washer register - checking ... 27

**Chapter 2. Crankshaft &
  conrods 29**
Connecting rods - checking 29
Crack testing 29
Straightness testing 30
Connecting rod bolts 30
Little end bushings 30

Big end bearing tunnel bore 32
Big end bearing crush 32
Crankshaft 33
Cleaning oilways 33
Grinding journals to optimum sizes . 34
Thrust washers 35

Chapter 3. Cylinder heads 37
Cylinder head choice 37
Exhaust ports (all head types) -
  modification procedure 38
Inlet port (all head types) -
  modification choices 41
Large inlet ports 42
Valve guides (all heads) - removal
  & replacement 42
Inlet port (B-type head) -
  modification procedure 44
Large inlet port 48
Inlet ports (straight port heads) -
  modification procedure 50
Large inlet port 52
Valve seats (all heads) - widths 54
Valves & seats (all heads) 55
Cylinder head (all types) -
  refacing 58

Tappet guides (all heads) -
  securing 58

**Chapter 4. Camshafts & valve
  springs 63**
Camshaft choice - road going 63
Camshaft choice - racing 64
Inlet valve closing 65
Exhaust valve opening 65
Valve overlap 65
High performance camshafts -
  sources 65
Some other considerations 66
Valve collets and keepers 67
Valve springs 67
Standard type 67
Stronger valve springs 68
Valves springs - checking 69
Valve spring retainer/oil seal -
  checking clearance 70
Camshafts - timing 71

Chapter 5. Ignition system 76
Contact breaker points 76
Vacuum advance 77
Static advance 77

SPEEDPRO SERIES

Centrifugal advance 77
Pre-ignition 77
Distributor modifications 77
Crankshaft damper - checking
  TDC & adding other timing
  marks 78
Finding top dead centre 78
Adding other timing marks 81
Static (idle speed) advance -
  setting 82
Ignition advance - checking 82
Ignition timing/total advance -
  finding optimum 83
Engine testing to find the optimum
  ignition timing 83
Summary 84

Chapter 6. SU carburettors 85
Twin 1³/₄ inch SU carburettors 86
Twin 2 inch SU carburettors 86
Spring loaded (swing) needle
  SUs 87
HIF7 SUs 88
Triple 2 inch SU carburettors 88
Modified SU carburettors 89
1³/₄ inch SU needles 89
2 inch SU needles 89
Conclusion 90

Chapter 7. Weber & Dellorto
  carburettors 92
Choke sizes & jetting 93
Tuning carburettors 94
Before starting the engine 94
Balancing airflow 94
Pump jets & inlet/discharge
  valves 94
Air correctors 94

Chapter 8. Exhaust system 96
Primary pipe length 96
Primary pipe diameter 98
Main pipe diameter 98
Silencers (mufflers) 98

Chapter 9. General information. 100
Camshafts 100
Tappet adjustment/valve stretch ... 100
Engine balance 100
Crankshaft thrust washers &
  clutch pressure plates 101
Summary 102
Clutch (driven) plates 102
Lightened flywheels 103
Timing chains 103
Oil pump 103
Sump (oil pan) 103
Oil cooler 105
Cooling system 105
Crankcase ventilation 109
Engine tune-up 109
Lucas fuel injection equipped
  engines 111
2.4 and 2.8 litre XK engines 111

Index **126**

www.velocebooks.com/www.veloce.co.uk
All books in print • New books • Special offers • Gift vouchers • Forum

Thanks & introduction

THANKS

Many thanks to Peter and Paul Cooper of COOPERCRAFT (specialists in restoration, sales and maintenance of Jaguars) in Westwood, Nr. Broadclyst, Exeter, Devon, England (e-mail: coopercraft@eclipse.co.uk) for being most helpful in allowing me to go through their vast range of engine parts and select various items for photographic purposes.

INTRODUCTION

This book is concerned with getting as much power as possible from the 3.4, 3.8 and 4.2-litre Jaguar XK engines, using readily available replacement parts and tuning components (all engine parts are still available). Although these engines have been around in one form or another for several decades, and are now out of production, they'll continue in use for many years to come (largely due to the historic nature of the cars they power).

Development of the XK engines continued over many years of production, resulting in exceptional power output in relation to engine size. Endurance racing in the early-1950s (Le Mans) meant that maximum reliability was required from XK engines and this was reflected in the modifications carried out by the factory and the rpm limit employed (Jaguar engineers never tested the C and D-Type engines above 5750rpm).

There were a number of overheating problems with early XK engines, culminating in the dramatic failures of all the C-Type cars at Le Mans in 1952. To ascertain the cause of the problem, one of the spare Le Mans engines, which hadn't failed, was tested on a dyno back at the factory, and the cylinder head was fitted was fitted with around 30 thermocouples to find out exactly what was going on inside the engine. The test revealed that the coolant was actually boiling iside the head, even though the flow of coolant through the head was normal.

The problem was traced to the fact that the ribbing inside the cylinder head casting trapped hot water in the top of the cylinder head. Although the coolant was still flowing through the cylinder head in the usual manner, not all of it was. A 100% flow of coolant through the cylinder head is required for optimum cooling. The problem had not presented itself in road going engines because they never had to deal with the amount of heat that was being generated during racing. Furthermore, the problem didn't appear during dyno testing because there was always an abundance of cool water to draw on. The ribbing inside the cylinder heads was turned 90 degrees allowing a clear path for the coolant across it. With this fixed, there were no more overheating problems.

The engines used by Jaguar were built with absolute reliability in mind, and this meant, as far as the likes of Jack Emerson were concerned, that the revs used had to be sensible and

SPEEDPRO SERIES

D-Type valve train components. The valve springs were triples.

This picture of a D-Type cylinder head combustion chamber shows the shape and size of the valves. Although similar to those on a standard Jaguar engine, they are different.

sustainable. The resulting engine was quite remarkable because it was still very similar to the production engines, and pushed the cars to just under 180mph. The main rival of the day at Le Mans was Mercedes-Benz, and its cars may well have gone slightly better with smaller capacity engines but, the straight eight masterpieces bore no resemblance to the production car engines; they were pure racing engines and the best available in the day. The Jaguar effort was fantastic by any standard.

From the late-sixties, a number of companies took this development to

The spark plugs are angled into the combustion chamber on the D-Type cylinder heads, whereas, the spark plugs on standard Jaguar heads are straight.

D-Type cylinder heads have round exhaust ports, as opposed to the usual rectangular ones.

its logical conclusion. This later work saw XK engines fitted with specially made components, such as super strong lightweight connecting rods and lightweight forged pistons, which allowed these long stroke engines to rev to 8000rpm with reasonable reliability. Such engines were usually equipped with triple Weber/Dellorto sidedraught or modified triple SU carburettors.

Today, most XK engines won't be taken to the ultimate tuning level because of the costs involved. Instead, they'll be modified using as many standard replacement parts as possible, including standard cast pistons which will give 8:1 or 9:1 compression. You should note, however, that if 9:1 compression pistons are used, and the block and the cylinder head planed to increase the compression ratio (CR), the CR should not exceed 9.5:1 (and even then, the octane rating and the quality of the fuel must be suitable for this CR). Retarding the ignition to stop an engine with too much compression 'pinking'/'pinging' during hard acceleration with heavy load is definitely wrong for XK engines: correct compression and correct ignition advance are essential.

Using cast pistons means limiting revs to 5500-6000rpm (depending on the type of pistons used). However, a very good and powerful engine can be built using standard replacement parts, including cast pistons, in combination with non-standard aftermarket specialist parts such as camshafts.

Although the design is old, these

SPEEDPRO SERIES

On D-Type cylinder heads, the stud bosses are different heights side to side.

The D-Type crankshaft is polished all over, as are the main caps.

D-Type piston and connecting rod. The connecting rod is highly polished all over. This was a job for apprentices.

engines have much to recommend them. They're excellent producers of torque and, in fact, very few modern standard production engines of similar capacity can match them for torque from such low rpm and over such a wide rpm band. This aspect of the XK's performance is due to long stroke, long connecting rods and good volumetric efficiency. In fact, this design trend has returned in recent years (note the Ford

SPEEDPRO SERIES

D-Type lightweight flywheel and clutch arrangement. Triple plate clutch was by Borg & Beck.

DCO3 Weber carburettors as used on the D-Type engines.

Weber carburettor and inlet manifold configuration for the D-Type.

V8 Triton engine, for example). These engines do everything very well and, so long as a sensible rpm limit is used, they don't break or wear out quickly.

The main performance improvement for XK engines comes from cylinder head modifications. The short block assembly is ready for many high performance applications just as it came from the factory.

Caution! These engines are unforgiving of any compromises and you must not cut corners during their preparation. If a modified XK engine does fail, it's likely to be because, in an effort to cut costs, a part wasn't replaced when it should've been. Once the correct components have been replaced, these engines get a new lease of life and faithful service will be your reward.

In the appropriate racing classes, XK engines are still a force to be reckoned with and the modifications detailed in this book allow major improvements in engine efficiency and power to suit all applications, though it has to be said that fuel economy will suffer.

Des Hammill

Using this book & essential information

USING THIS BOOK
Throughout this book the text assumes that you, or your contractor, will have a workshop manual specific to your engine to follow for complete detail on dismantling, reassembly, adjustment procedure, clearances, torque figures, etc. This book's default is the standard manufacturer's specification for your XK engine so, if a procedure is not described, a measurement not given, a torque figure ignored, you can assume that the standard manufacturer's procedure or specification for your engine needs to be used.

You'll find it helpful to read the whole book before you start work or give instructions to your contractor. This is because a modification or change in specification in one area may cause the need for changes in other areas. Get the whole picture so that you can finalize specification and component requirements as far as is possible before any work begins.

For those wishing to have even more information the following Veloce titles are recommended further reading: *How To Build & Power Tune Distributor-type Ignition Systems*, *How To Build & Power Tune Weber & Dellorto DCOE & DHLA Carburetors*, *How To Build, Modify & Power Tune Cylinder Heads*, *How to Choose Camshafts & Time them for Maximum Power* and *How to Modify & Power Tune SU Carburettors*.

ESSENTIAL INFORMATION
This book contains information on practical procedures; however, this information is intended only for those with the qualifications, experience, tools and facilities to carry out the work in safety and with appropriately high levels of skill. Whenever working on a car or component, remember that your personal safety must ALWAYS be your FIRST consideration. The publisher, author, editors and retailer of this book cannot accept any responsibility for personal injury or mechanical damage which results from using this book, even if caused by errors or omissions in the information given. If this disclaimer is unacceptable to you, please return the pristine book to your retailer who will refund the purchase price.

In the text of this book "Warning!" means that a procedure could cause personal injury and "**Caution!**" that there is danger of mechanical damage if appropriate care is not taken. However, be aware that we cannot foresee every possibility of danger in every circumstance.

Please note that changing component specification by modification is likely to void warranties and also to absolve manufacturers of any responsibility in the event of component failure and the consequences of such failure.

Increasing the engine's power will place additional stress on engine components and on the car's complete driveline: this may reduce service life and increase the frequency of breakdown. An increase in engine power, and therefore the vehicle's performance, will mean that your vehicle's braking and suspension systems will need to be kept in perfect condition and uprated as appropriate. It is also usually necessary to inform the vehicle's insurers of any changes to the vehicle's specification.

The importance of cleaning a component thoroughly before working on it cannot be overstressed. Always keep your working area and tools as clean as possible. Whatever specialist cleaning fluid or other chemicals you use, be sure to follow - completely - the manufacturer's instructions and if you are using petrol (gasoline) or paraffin (kerosene) to clean parts, take every precaution necessary to protect your body and to avoid all risk of fire.

Chapter 1
Cylinder block

There have been several versions of cylinder block for the bigger capacity XK engines; the 4.2 block having the most of variants and, in certain forms, the greatest number of problems.

Caution! - Although the overall strength of XK blocks is beyond question, most are old now and will have covered a considerable mileage. When ANY of these engine blocks is to be built up, whether for high performance road use or competition use, certain things simply have to be checked as detailed in this chapter. The XK engine is not a forgiving engine when it comes to assembly criteria, and will fail prematurely if vital checks are not made. To avoid future problems, a number of components and tolerances have to be checked and faults remedied if they are found.

It's advisable to read right through this chapter so that you know what to look for when assessing the suitability of a block for high performance use.

GENERAL CHECKS & ADVICE

Caution! - Before starting work on an XK engine you must make sure that each cylinder bore has been 100% crack tested by an engine reconditioner/engine machine shop using the Magnaflux process. In fact, Jaguar blocks are very rigid structures and cracking is not normally a problem. The cylinder sleeves in sleeved blocks will sometimes be cracked, though even this problem is not common.

Caution! - The main caps should also be crack tested by the Magnaflux process. Purchase new main cap bolts to use in the rebuild.

Caution! - It is essential that the water jacket of the block is thoroughly cleaned to remove all traces of the silt which may have accumulated over the years. The water chambers of the rearmost cylinders are very often completely full of silt (because the engine is inclined when installed in road cars).

Cleaning the water jacket area can be quite a difficult job and requires removal of all the freeze (Welch) plugs. Using welding rods, or similar, poke all around the water jacket voids and make sure that all is completely clean and clear. Even if the cylinder block has been 'hot tanked,' (chemically cleaned) check the water jacket area.

If you are going to rebore the cylinders because of damage or wear, only go to the very next over-size possible. Boring to the maximum size just to gain a few extra ccs is not a good idea: all it does is remove a life or two from the block with no measurable difference in power output being gained. Some of these blocks are also getting expensive to buy now: for example, the 3.8-litre block, if you can find one, commands a premium price. Making XK blocks last is now an important priority!

Standard replacement type piston and ring sets are available from Jaguar and a number of aftermarket

CYLINDER BLOCK

companies. All types are made to original equipment (OE) specifications. Pistons are readily available in 0.010, 0.020, 0.030 and 0.040 inch oversizes. 0.060 inch plus sizes are available for some engine blocks, but they're not common.

Many replacement cast pistons feature split skirts which limit the rpm capability to 5500rpm. Solid skirt cast pistons are safe to 6000rpm. If an engine is going to be used exclusively for racing purposes, and 5500-6500rpm seen on the rev counter all of the time, forged pistons will be required to ensure absolute reliability. While it's quite possible to use cast pistons in 5500-6000rpm applications, and replace them at regular intervals (10-20 racing hours) there comes a crossover point on the basis of cost versus the near guaranteed reliability of forged pistons.

There is no substitute for perfectly round parallel bores. New pistons and rings fitted to perfect bores with the correct piston to bore clearance is the ideal. If there is any appreciable bore wear (more than 0.001in/0.0254mm) consider the bores too worn for high performance use. Fortunately, bore wear is not a significant problem on Jaguar engines due to the fact that Jaguar always used good quality materials for its engine blocks and liners. It is not uncommon to find a Jaguar engine, which has done well over 100,000 miles, with less than 0.001 inch of wear in the bore, and even visible hone marks lower down.

Always check the bore size, or have it checked, with an inside micrometer: just because a bore looks unworn does not mean it is. Bores get honed to clean them up, and often the end result is a bore that looks alright but isn't.

On average, standard recommended piston to

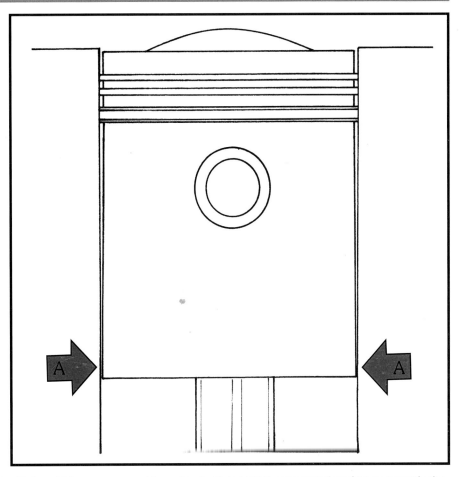

Piston skirts are measured from A to A. Piston skirts are tapered as they go towards the piston ring grooves.

bore clearances are 0.0007-0.0013in/0.01778-0.0330mm (there is some variation here). This, however, is just too tight for a high performance engine and seizure is possible.
Caution! - Take it that the minimum piston to bore clearance to have on any of these engines is 0.0027in/0.0685mm and that the maximum, using standard type replacement cast pistons, is 0.0032in/0.0812mm.

If the piston to bore clearance is too much the pistons won't seize, but there is the problem of the piston having too much sideways movement in the bore. This will cause the piston rings to wear out very quickly (become barrel sided) and there will be much more blow-by than normal, resulting in crankcase pressurisation.

Pistons, of course, wear in several places and, once certain critical measurements (sizes) are lost through wear, replacing the pistons is the only sensible solution. Once the skirts, for example, are no longer on size (maximum wear 0.001in/0.0254mm) the pistons are no longer serviceable for a high performance engine. For example, if the piston to bore clearance when new was 0.0028 inches, and the pistons are now worn by 0.0017 inches, the piston to bore clearance has become 0.0045 inches/0.114mm,

SPEEDPRO SERIES

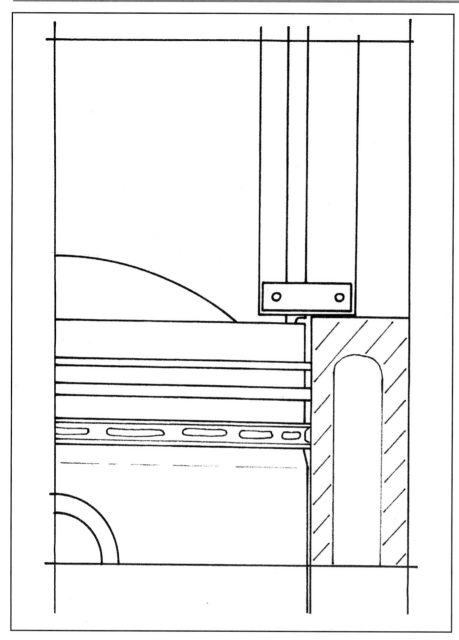

The tail of a vernier calliper is used to measure the distance from the top of the block to the top of the piston crown when the pistons are at top dead centre (TDC). Both the top of the block and the piston crown must be clean.

These two measurements need to be recorded for future reference, especially if the compression ratio is to be increased. Pistons that are removed from the engine should not be thrown away. The reason for this is that there is often some difference in piston crown height between Jaguar pistons and replacement part manufacturers' pistons. The new pistons need to be compared to those removed to see if the crown heights are different and, if they are, by how much. This information can be used later when calculating how much material can be removed from the top of the block if the compression ratio is going to be increased (i.e. making the piston crowns flush with the top of the block deck surface).

As a general rule, if you need to replace the rings, the pistons should also be replaced. However, for any engine used in a competition environment the rings almost always get replaced whenever the engine is stripped down (usually once a year). If the rings are replaced very frequently the pistons might well be still on original size and not need replacing. Keep a record of the original piston skirt sizes and compare them with subsequent measurements. If the rings are replaced often, take piston skirt wear as an indicator of when it's time to replace the pistons.

Something that is not often taken into consideration is the fact that metal fragments lodge in the softer materials of components used in engines, such as piston skirts and bearing surfaces. Once particles are in the surface of a piston skirt, for example, the bore wall will wear. Replacing the pistons at sensible intervals prevents this happening.

Once the top ring groove has more than 0.001in/0.0254mm of vertical wear in it, the piston is past

which is too much. Piston skirts are measured for wear with an outside micrometer.

Before removing any of the pistons from the block, each piston should be positioned at top dead centre (TDC) and the distance from the block's deck down to the piston crown measured. The compressed thickness of the cylinder head gasket should also be measured (vernier calliper measurement acceptable).

CYLINDER BLOCK

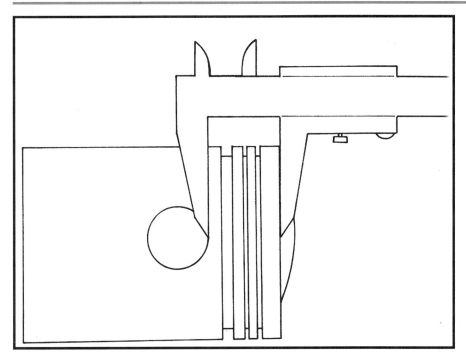

Vernier calliper being used to measure the gudgeon (piston/wrist) pin to piston crown height.

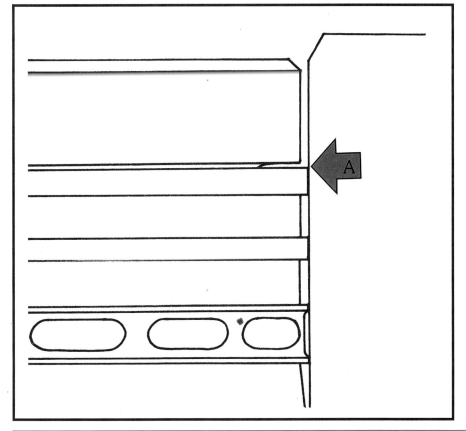

The top piston ring groove wears above the ring, at A.

SPEEDPRO SERIES

its best and needs replacement (a new ring has virtually no noticeable vertical movement when fitted to the top ring groove). With a worn top ring groove, top ring sealing will not be efficient and, although an engine in this condition will still run well enough, it will not run as well as it could and performance will reduce quickly.

If you have a set of pistons which are completely 'on size' apart from the top piston ring groove, you can get thicker than standard piston rings made. Johnson Piston Rings, 918 Great South Road, Penrose, Auckland, New Zealand, makers of 'Indianapolis 500' custom made piston rings, will supply you with six, top quality material, top compression rings which are 0.001in to 0.010in thicker (or whatever is required) than standard for a very reasonable price. They will also package and freight these components to anywhere in the world for a very reasonable cost. If your top ring grooves are worn by 0.003in, for example, you can order a set of 0.005in thicker top/first compression piston rings, and then have the top edge of the top ring groove (the worn one) machined square in a lathe with 0.001in clearance on the new oversized ring. This will give your pistons a new lease of life, as the ring to piston side clearances will be back as per standard. Johnson Piston Rings will, of course, supply the rest of the piston ring set if required. You can also order a few spares of each ring size if you want to for little extra cost (just in case you happen to break one). All over-sizes are available.

This company have been in business for years doing this sort of work and can be considered experts in this field. They are contactable on +64 9 579 7219 or Fax +64 9 579 8788, or write to Johnson Piston Rings, PO Box 12230, Penrose, Auckland, New Zealand. The front desk staff are very experienced and all that is required is to state the make and model of engine, the year of manufacture, the size of the pistons (plus 0.010in/0.25mm, 0.020in/0.50mm etc.) and how thick you want the top ring to be (in thousandths of an inch/mm). Have the top ring grooves of the pistons machined by an experienced engineer once you receive the new rings. The surface finish needs to be as smooth as possible. Most engine reconditioning companies or precision engineering workshops will do this work for you at reasonable cost. The clearance between the ring and the piston groove is 0.0007-0.001in/0.0177-0.0254mm.

Caution! - If the piston (gudgeon) pin circlip grooves show signs of wear - for example, when the circlip can be easily rotated in the groove and the actual groove in the aluminium of the piston shows signs of burring - the piston is past its best and must be replaced to avoid failure (a circlip coming out, for example).

Caution! Never re-use circlips, and only ever fit circlips to a piston once. Whenever circlips are fitted into pistons, squeeze the circlips down just enough to get them into the gudgeon pin bore. Always check to see that standard type circlips are fitted into a piston with the sharp edge facing away from the end of the piston pin. Circlips are stamped out of steel strip on a press and, as a result of this manufacturing process, the sides are not as parallel as they might appear. If a circlip comes out of the groove in the piston the engine will usually end up being seriously damaged.

Because the standard circlips have been known to come out of their grooves for no apparent reason, many people have successfully used Teflon buttons for gudgeon pin retention. With Teflon buttons fitted, the engine will have to be inspected periodically because the Teflon wears away with use. Failure to replace worn Teflon buttons means a ruined engine because the gudgeon pin wears the bore wall away.

A further possibility is to retain the circlips, but have Teflon buttons made that fit over them. The Teflon buttons prevent the circlip coming out of its groove and will prevent failure. The circlip is sandwiched between the end of the gudgeon pin and the Teflon button which acts as a fail safe if the circlip fails.

It's a good idea to buy the pistons before boring out the block. As a result, you can have your engine reconditioner measure each piston and, if necessary, bore and hone each bore to suit each individual piston skirt size. Most pistons will be identical in size (to a few tenths of one thousandth of an inch). Add the piston to bore clearance to each piston skirt size and take pains to make sure that all of the pistons end up with an identical piston to bore clearance.

When the pistons you intend to use are fitted on to their respective connecting rods, their alignment should be checked by an engine reconditioner/engine machine shop using special equipment. This is done on the same jig that is used to check the connecting rod alignment but with a supplementary fixture to hold and align the piston. The connecting rods are all checked first, to make sure that they are in correct alignment, and then the pistons are fitted and the overall piston/connecting rod alignment is checked. When engine components are checked in this manner there are usually very few assembly or running problems later on. Although it is quite rare to find a piston that has not been accurately machined, it does happen from time to time, and this test will

CYLINDER BLOCK

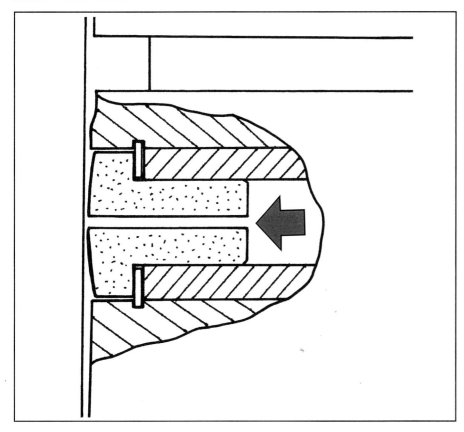

The circlip is fitted into the groove in the piston's gudgeon pin bore in the normal way. The Teflon button's stem fits into the bore of the gudgeon pin, and the circlip is effectively trapped in its fitted position. It's also possible to use Teflon buttons without circlips (see text). Note the 1/16in hole through the centre of the Teflon button (arrowed), which is to allow, heated and expanded air to escape.

find an incorrectly machined piston. When this happens, the piston can be exchanged for one that is correctly made.

Have the top surface of the block remachined when the engine is stripped down for rebuild. This is necessary, in the first instance, to prevent cylinder head gasket failure. Secondly, the removal of material from this surface is a most satisfactory way of increasing the engine compression. In many instances, the gap between the top of the piston crown and the top of the block is 0.040 inches/1.0mm, or even more, and this amount of material can safely be removed from the block to effectively bring the top of the piston up flush with the top of the cylinder block.

Caution! - Since it forms part of the cylinder head gasket surface, the timing chain cover must be fitted to the block when the block's deck is machined.

Once a block has been planed to increase the compression ratio of the engine, the clearance between the crown of the piston and the combustion chamber edges at top dead centre (TDC) must be checked. Failure to do so might result in the piston hitting the cylinder head. There needs to be a minimum clearance of 0.040 inches/1.0mm between the edge of the piston crown and the edge of the cylinder head at TDC. Most pistons feature chamfered edges which prevent contact, but the clearance still needs to be checked if a lot of material has been removed from the block's deck surface.

The larger the piston to bore clearance, the more the piston will 'rock' in the bore at top dead centre (tdc) and the tilted edge of the piston will be higher in the bore. The higher the rpm being used, the more the connecting rods will stretch and lift the piston in the bore. If higher rpm than 6000 is being used an extra 0.010in/0.25mm of clearance will have to be given to avoid piston crown to cylinder head contact.

Caution! - To facilitate perfect ring seating, the cylinder bores should always be 'power honed' on a machine specially made for the purpose. Through its mechanical action, such a machine will ensure that the honing pattern is absolutely correct (correct cross hatch angle). Most, but not all, engine reconditioners/engine machine shops have this sort of equipment. Jaguar XK engines have very long strokes and, as a consequence, long bores. When these blocks are hand honed there is a possibility that the honing pattern will be incorrect, leading to poor ring seating and an engine that burns oil.

After the bores are finished and cleaned, the rings need to be 'check fitted' into the bores and have their end gaps checked. **Caution!** - Although new rings are usually correctly gapped, errors can still be found: avoid problems by checking every ring in the cylinder bore in which it is going to run.

For the top compression ring, the minimum recommended end gap is 0.015in/0.381mm (never less) and avoid fitting rings that have more than 0.020in/0.508mm end gaps.

SPEEDPRO SERIES

The minimum end gap for the second compression ring is 0.012in/0.304mm, maximum 0.015in/0.381mm. Oil control rings are less critical and are rarely the minimum size (0.015in/0.381mm). Avoid fitting oil control rings which have end gaps of more than 0.040in/1.016mm.

Rings that have too small an end gap will butt when the engine gets warm, resulting in engine failure. Rings which have large end gaps give a poor gas seal.

While it's quite normal to hone the bores to conventional specifications, there is a further operation that many race engineers like to use to improve ring seating/sealing. This procedure is called 'plateau honing'. It's a common practice in automotive engine reconditioning circles, with many workshops doing this to all of their engines. The bores are honed in the normal manner using 200 grit stones, then 500 grit stones and then 'corkings' are used (Sunnen, Delapena, etc., being suppliers of such equipment). The 'corkings' dress the tops of the honing marks, or round them off if you like, which gives the bores a polished look.

With Total Seal gapless rings fitted, for example, the percentage of cylinder 'leak down' can be virtually zero, as opposed to 2-3% if those rings had been fitted into conventionally honed bores. With normal gap type rings, the percentage of 'leak down' can be reduced to 2-3%, as opposed to 6-8%. It's a significant reduction in leakage past the rings.

The only drawback with 'plateau honing' cylinder bores involves making sure that the rings seat. Because the bore wall surface is semi-polished, the tendency is for the rings to glaze if the engine is not subjected to loading more or less immediately. This isn't really a problem if the engine is first run on an engine dyno or rolling road, and loaded within about 20 seconds of starting. Bore glazing has put people off 'plateau honing' because the only thing you can do in this situation is strip the engine, re-hone the bores and fit a new set of rings.

One thing you can do to help the rings seat/seal immediately is to use CRC or WD-40 penetrating oil as a lubricating medium on the bores, rings and pistons, instead of engine oil. The bores, pistons and rings can be sprayed liberally with penetrating oil since only so much will stay on the surfaces, the rest will drip off. Using penetrating oil, coupled with loading the engine as soon as possible, causes the rings to start the 'bedding in' process immediately. In fact, even if the engine is not subjected to load immediately after starting, it is highly likely that the rings will still bed in satisfactorily through the use of the penetrating oil alone.

Some time ago I had to replace a cracked block in between race days. A new block was prepared and the bores 'plateau honed'. The damaged bearings, crankshaft, pistons, connecting rods and the old, well-used rings (Total Seal gapless rings) were all fitted back into the new engine block. The rings were plus 0.060in ones, and we weren't able to source new ones in time. With the engine assembled with CRC penetrating oil on the bores, pistons and rings, it was started on a rolling road dyno and given some loading after about 20 seconds. After twenty minutes of running, a 'leak down' test was done on all cylinders (air pressure was applied to each cylinder and the leakage past the rings expressed as a percentage read off). All cylinders were reading 0.5 to 1%. A run was done to maximum rpm and a maximum power reading taken. It was within 0.5% of what it had been before the block failure. The engine went on to complete the rest of the racing season without needing to be stripped down again, and the oil used for racing purposes was in the engine from the time it was started up. The old and more than half worn rings shouldn't have seated as well as they did!

COMPRESSION RATIO (CR)

It's worth noting that most standard replacement cast pistons on the market these days give an 8:1 compression ratio for unmodified XK engines. In many instances, this is lower than the original CR. It's fair to say that 8:1 is a bit low, though, as it happens, XK engines are not very compression sensitive (meaning they go very well even if the engine has a low compression). However, what with block deck and cylinder head planing, and substituting a thick cylinder head gasket for a thin cylinder head gasket, the compression ratio of XK engines can be increased to 8.5:1 through to 9:1, and more, even if the pistons fitted are essentially 8:1 CR units.

9:1 compression pistons are available from some replacement piston companies for some XK engines. With block and cylinder head planning the compression can be got up to 10:1 with these pistons. With the trend these days for lower compressions and unleaded fuels, however, 10:1 compression is the 'ragged edge,' and it's just not necessary to go to great lengths to build as much compression into the engine as possible. By all means go this high if getting the absolute maximum power and torque out of the engine is essential, but be aware of the fact that this could lead to an engine failure. Cast pistons are only so strong and, if the engine is suffering from pre-ignition at times, could damage the engine if they fail.

CYLINDER BLOCK

Caution! Resist the temptation to go too high with the compression ratio. Compression is not the 'be all and end all' with these engines, although there is no doubt that they do go better with high compression as opposed to low compression. Exceed the limit, though, and it is likely that the engine will end up damaged.

With such a vast variation of fuel qualities available around the world it's difficult to give hard and fast rules with regard to compression ratios. However, you would be safe to assume that on the low side, 8:1 compression is almost always suitable for 90 to 97 octane fuel and that 8.5:1 compression is almost always suitable for use with 95 to 97 octane fuels.

No guarantees, but usually 8.5:1 compression should be okay with 90 octane fuel, 9.2:1 compression should be okay with 95 octane fuel and 9.5:1 compression should be okay with 97 octane fuel: however, thorough testing to make sure that all is well is essential. 10:1 compression has been used successfully with good quality 97 octane fuel, but this is most definitely the upper limit and quite risky.

Although there is some leeway with compression ratios, to run a 9.5:1 compression engine on 90 octane fuel in a high load situation is just asking for trouble. By all means try an engine with the compression on the high side but be prepared to reduce the compression if it proves necessary. This, of course, will involve some reworking, which, because of the complexities of these engines, most people will naturally want to avoid. Making a sensible decision on the compression ratio to be used in the first place is the best policy.

Cylinder head gaskets of different thicknesses are available from the various gasket companies. If the compression ratio has been taken too high, the cylinder head gasket can be changed for a thicker one which may well be enough to reduce CR by a suitable amount. It is a good idea to check whether or not different thickness cylinder head gaskets are available for your particular engine before the engine's compression is increased to a high ratio. If the compression does end up too high, changing the cylinder head gasket could be a simple solution. Do the numbers first so that you will know if this option is available to you. To make this sort of change a viable option, the cylinder head gaskets available will need to vary by at least 0.020in/ 0.508mm in thickness.

The original compression ratios of standard engines were never very high. The range of CR, from the earliest to the latest engines, was 7:1 to 9:1. The highest amount of compression used on the racing engines in days gone by was in the range 10.5 to 11:1, and high octane leaded fuel was definitely used. XK engines will, however, go very well with CRs of between 8:1 and 9:1. Even a 7:1 compression engine (with flat topped pistons) will go well. With a correctly modified cylinder head fitted, good induction and exhaust systems, and the engine tuned properly, the CR does not seem to be of too much importance (except when it is too high for the octane rating of the fuel being used ...).

The reason why XK engines can run well on what might appear to be quite low CRs, is that they have very long strokes, very long connecting rods and, in modified cylinder head form, very good volumetric efficiency (cylinder filling). They don't need to have a lot of compression to generate good engine torque and produce good engine power. With these engines volumetric efficiency is more important than CR.

PETROL/GASOLINE FOR JAGUAR XK ENGINES

Although the majority of fuels commercially available around the developed world these days are unleaded, many countries are still using leaded fuels (tetraethyl lead added to increase the octane rating). The current unleaded fuels available are the same basic fuels as before, but some now have their octanes boosted through the use of aromatic hydrocarbons, iso-octanes, or alkanes, as opposed to tetraethyl lead. The lower octane unleaded fuels (91 RON, for example), are much the same as they always were but with the tetraethyl lead removed. These fuels didn't need additional substances to boost the octane once the tetraethyl lead had been removed, but rather, the existing levels of the constituents were increased to bring them up to the required octane requirement.

The two octane ratings of petrol/ gasoline are the Research Octane Number (RON) and the Motor Octane Number (MON). In the late 1920s, the Co-operative Fuels Research Council (CFR) in the USA devised a method for determining the octane ratings of fuels. This involved developing a special single cylinder engine with variable compression ratio and ignition timing. This engine, commonly referred to as a CFR engine, is still used to test samples of fuel for RON and MON. However, before any fuel sample is tested, the CFR engine is calibrated using a pure chemical mix 'reference fuel' which, because of its specific chemical content, is guaranteed to be 100 octane.

To determine the RON octane rating, the CFR engine is run at 600rpm with a set amount of ignition timing as prescribed by ASTM D2699/IP237 (13 degrees before top dead centre). This rating is regarded as being representative of how the particular fuel will cause an engine to go at start up and idle. ASTM stands for American Society for Testing and Materials and D2699 is the criteria for its RON test. IP stands for Institute of Petroleum and 237 is its number for this testing regime (same test exactly).

The method of obtaining a MON octane rating is conducted under ASTM D2700/IP236 criteria. The CFR engine is run at 900rpm, the compression ratio is increased, and the ignition timing is advanced. The octane derived from this test is regarded as being representative of how the particular fuel sample will cause an engine to go on the road at cruise conditions or motorway driving.

The UK has 95 octane (RON) Premium unleaded, 97 octane (RON) Super unleaded, and Shell Optimax, which is 98.3 octane (RON). Shell Optimax is the highest octane fuel that is readily available in the UK, at the time of writing, and is ideal fuel for use in a high compression engine. The use of 10.0:1 is possible for racing purposes with this fuel. For road use, and where low rpm use under high load is required, consider 9.0-9.5:1 to be the range to consider using.

Most developed countries around the world use unleaded fuels with octane ratings of between 91-97 (RON). High octane (RON) tetraethyl leaded fuels are still available for off-road racing purposes in many countries, and, if you intend to only use these sorts of fuels, you can build any known usable compression ratio (11.0:1, for example) into your XK engine, with the full knowledge that the fuel will be compatible. Some countries still offer 'Avgas', packaged as automotive racing fuel for racing use, even though the country as a whole has changed over to unleaded fuel for all road-going cars. Avgas is 107 RON and 100 MON, and has 0.85 grams of tetraethyl lead in it per litre. There is a low lead version of Avgas, which has 0.75 grams of tetraethyl lead in it per litre. For all general racing use, both fuels are fine. Tetraethyl lead, which is a really nasty substance, is apparently only manufactured in one plant in the world, and that is located in Russia.

The Avgas of today used to be known as 100/130 in days gone by (in the 1940s, 1950s and 1960s). The 100 numbers are the MON rating, while the 130 was a rating obtained from a supercharge test method. The supercharge rating became less applicable as time went by, as the use of very large supercharged piston aircraft engines declined. The RON rating was 107 for the 100/130 aviation fuel.

The USA has taken the RON and MON octane rating system one step further by introducing an Anti-Knock Index number (AKI), which is based on the RON and the MON added together and then divided by 2. As you can see, when you are talking octanes, you need to be quite clear what criteria are under discussion. If you look at the USA's Sunoco GT100 street legal unleaded fuel, for example, which is advertised as being 100 octane, this fuel has a 105 RON and a 95 MON making it extremely good fuel. The 100 octane rating you see advertised is the AKI, which is 105 + 95 divided by 2 = 100. This fuel will run any naturally aspirated engine with up to a 12.0:1 compression ratio in most instances. VP's Performance Unleaded is an equivalent fuel.

Note. Engines which have cast iron valve seat inserts fitted to them have to use a lead substitute additive with the fuel to prevent valve seat recession (the excellent TetraBoost, for example, the details of which can be seen at www.tetraboost.com). With the knowledge of the RON or AKI of the fuel available for your engine, a decision can be made as to what compression can be used.

In the USA, there are three basic grades of street legal unleaded fuel on sale, plus high-octane unleaded fuel. For the purposes of matching the fuel available to a compression ratio that will be suitable, use the RON ratings of the US fuels listed in the accompanying octane versus compression ratio chart.

Regular Unleaded 87 octane is 91 RON - 83 MON
Mid-grade Unleaded 89 octane is 94 RON - 84 MON
Premium Unleaded 91 octane is 96 RON - 86 MON; 92 octane is 97 RON - 87 MON; 93 octane is 98 RON - 88 MON

The higher the octane rating of the fuel available, the higher the compression ratio that can be built into the engine and successfully run (without pinking). The higher the compression ratio used the better. It's a mistake, however, to run a compression ratio that is higher than the fuel can cope with. The engine just won't respond as it should, and can't be driven hard. It's better to have 9.0:1 built into an engine and be able to use wide open throttle under all circumstances, than it is to have 10.0:1 and have to be selective about when wide open throttle can be used to avoid pinking.

Different countries have different companies' products on sale, and while many companies use the same oil refinery, many do not. There can be differences in the fuels available of the same octane rating. The RON rating listed at the pumps will usually list the minimum RON octane ratings that the country requires. What is not always apparent is that while many fuels meet

CYLINDER BLOCK

the minimum specification, some products are above this. For example, if a minimum specification listing is 91 RON - 81 MON, some fuels could well be 93 RON - 82 MON. For the most part, these high octane fuels will cause your engine to run better and return more miles to the gallon/kilometres per litre. The end result of this is that you do need to check the fuels of the same octane rating in your area just to see if there are differences.

The way to do this is to fill the tank to the brim, record the odometer reading, and then drive the car around town for 50 miles/80km and then fill the tank to the brim again. Record the distance travelled and the fuel used. Next, drive the car on a motorway for a similar distance and then fill the tank to the brim again. Record the distance travelled and the fuel used. The miles per gallon/ kilometres per litre of each type of driving can be easily calculated and an average calculated. While this testing regime may seem to be a bit unscientific, it is very effective, and accurate enough for most. Shorter distances and the same piece of road offers quick and slightly more accurate data readings.

Caution! - Unleaded fuels don't have the 'scavengers' and lubricating qualities of the tetraethyl leaded fuels of years gone by, and 'valve sticking' can now a problem. The solution is to have all of the exhaust valve guides K-Lined. Most engine reconditioning workshops use this system nowadays, and the cost per guide is reasonable. Another way of preventing valve sticking is to have the valves Tuftrided.

3.4-LITRE CYLINDER BLOCK - NOTES

The 3.4-litre engine block is the most rigid XK block of all. As a rule, it doesn't have factory fitted cylinder sleeves and, consequently, has thick cylinder walls; if you do find that your 3.4 block has been sleeved this is not normally cause for concern. The advantage of an unsleeved block is that it can be bored out to a maximum of 0.060 inches oversize and still remain very strong. If you have the choice, an unsleeved block is the best option.

The only disadvantage of the 3.4 block is that the engine capacity is the smallest of the three available. In spite of this obvious factor, do not be deterred from using a 3.4-litre engine since it's a 'rock solid' block (which is more than can be said for some of the later ones on the basis of head stud threads). Also, a capacity of 3.4-litres doesn't mean that the will lack power, this is still a decent amount of displacement for high performance.

3.8-LITRE CYLINDER BLOCK - NOTES

The 3.8-litre blocks were all factory sleeved. This, in itself, will not cause any problems so long as you do not bore the cylinders any larger than is absolutely necessary. Aim to restore the bore to the very next oversize piston size possible. As mentioned earlier, boring out XK blocks to the maximum possible to gain a few ccs is just not a good idea. All it does is remove possible lives from the block, without any gain in engine power, and it makes the liner walls thinner. Keeping the liner walls as thick as possible is desirable. Thin liners have a tendency, in some high performance situations, to crack lengthways resulting in a wrecked engine. To avoid problems, always observe Jaguar's maximum recommended over-bore size of plus 0.040 inches in factory-linered engines.

4.2-LITRE CYLINDER BLOCK - NOTES

These blocks were all factory sleeved and all had 'Siamesed' bores. Early Mark X triple carburettor engines and the early E-Type engines had what is termed a 'thick decked block.' They also have block deck mounted head studs (short studs) as opposed to the longer crankcase-mounted head studs (long studs) found on later engines. The early blocks are the best blocks if you can find one: they're the most rigid of the 4.2-litre engine blocks. Like the 3.8-litre blocks, only bore out to the very next oversize when restoring cylinders: keeping the liners as thick as possible is desirable. Maximum factory recommended oversize is plus 0.040 inches.

Long head stud blocks

A common problem with the later 4.2-litre blocks is that the long studs sometimes break deep down in the block. Also, the threads deep down in the crankcase area can become damaged. These long head stud blocks were alright when they were new, and perhaps even for the first ten years, but that's now a very long time ago. If you find a long stud engine with these problems don't be tempted to repair it, find another block in good condition instead. Engine blocks are in reasonably good supply at the moment and there is no reason to start repairing damaged ones.

Any long stud block should be tested for stud problems before any reconditioning work is done. Most of the problems start on these blocks when they were run for years without any rust inhibitor in the cooling system. The long studs get corroded leading to failure (the stud either breaks off or starts to stretch). The threads of the block and studs can also corrode and, even if the stud is removed in one piece, the threads are often damaged. On initial inspection, the threads may not appear to be damaged, and this

SPEEDPRO SERIES

is where a mistake can quite easily be made. In many instances, the stud will be removed and the block will have a lot of money spent on it only to find that when a new set of studs is fitted a particular stud is not a good fit in the threads of the block. In many instances the thread will strip out of the block when the cylinder head is torqued to correct tension.

If this problem is encountered during engine assembly, it can be solved by having a precision engineering works make up a special stud which has an oversized thread on it. This will mean that the particular stud is non-standard and is machined to suit the now incorrectly threaded hole in the block. This will mean that the thread is 0.005 inch to 0.010 inch larger in diameter than an original one.

Such threads are screw cut in a lathe, as opposed to being 'rolled' as standard threads are. However, providing the operator turning the stud knows what the stud is for, takes care to ensure that the stud is made out of top quality high tensile steel, and ensures that a small radius is 'put on' on the turning tool, the stud will generally not fail in service (break through having a stress point). Sharp corners are what cause studs to fail in this situation. Good machinists know this as they are dealing with similar problems on a daily basis. **Caution!** - The oversized stud solution will only work if the thread in the block is in 'good' condition, even if it is technically oversized. If the thread is damaged, with bits missing or crumbling away, the fitting of a specially made oversized stud will not work.

Wasting time and effort with a block that is basically unsuitable for further use can be avoided by checking the condition of the threaded holes immediately after the studs have been removed. New studs should be fitted into the block and the cylinder head put back on the block and the studs all torqued up to the recommended setting. This is nothing more than a dummy assembly to prove whether or not the threads in the block are good enough to take the proper bolt torque. If a stud doesn't take the correct torque it means that there is something wrong somewhere (stud, block or both).

As a further check, before the cylinder head is put on to the engine measure the height that each stud protrudes above the block's deck with a vernier calliper. After the cylinder head has been fitted, and the studs proved able to take the torque, the cylinder head is removed and then the distance that the studs protrude above the block deck is measured again. They should all be much the same as they were before the test. If any stud is noticeably different (higher), refit the cylinder head, torque the studs up again and then remove the cylinder head. Check the height of all of the studs again and this time the heights must not differ. If a stud keeps 'moving', the block is faulty and, while it may not fail immediately, premature failure will occur.

Do not invest time and money in a block that has a problem. Bad parts are worse than no parts at all and, if the block is wrong the whole engine is wrong. Some blocks, which have had a rust inhibitor in them all their lives, can be in perfect condition in all respects. Check, as much as possible, how clean the water passageways are in the block. Remove the freeze (Welch) plugs and see what the condition of the cast surfaces are. In many instances, if these surfaces are like new, the stud and block threads will also be in excellent order.

ALL SHORT STUD BLOCKS
While short stud blocks rarely have

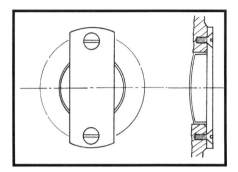

A sectioned side view of the block, freeze plug and strap. Use a sealant on the fastening screw threads.

block deck thread problems, these blocks can also be checked in a similar manner to long stud blocks (using a dummy assembly to check the integrity of each head stud and block thread).

FREEZE (WELCH) PLUGS - STRAPPING
The standard freeze plugs are almost always reliable in standard road going situations, but they've been known to come out when the engine is stressed (in competition, for example). To prevent this, the freeze plugs need to be 'strapped.' This precaution means that even if a freeze comes loose it will not be able to fall out completely and the engine won't loose all of its water in about 5 seconds and get ruined. If a strapped freeze plug leaks, but can't fall out, any water loss will be gradual. The early type of freeze plug is more prone to this problem than those in later blocks. Don't simply replace the freeze plugs and hope that they'll stay in place. It could cost you the engine! Periodic inspection of all freeze plugs to check for leaks is recommended, but there is no doubt that 'strapping' plugs is sensible in high performance applications.

MAIN BEARING CRUSH - CHECKING
A vital check that MUST be made

CYLINDER BLOCK

A main bearing tunnel being measured with an inside micrometer. Measurements should be taken in three places.

is that of 'main bearing crush' - a term that may well be unfamiliar to many people. This procedure involves checking that the new main bearing shells are fitted into the block correctly. The main bearing shells fit into an aperture called the main bearing tunnel, comprising the block and the main cap. The main bearing tunnels should be perfectly round. Fortunately, the XK Jaguar block is as solid as it looks and there are usually few problems here. Each main bearing cap combination must be checked with an inside micrometer to make sure that the aperture is in fact round and 'on size.' (within specification). The minimum diameter of the main bearing tunnels is 2.9165in/74.079mm and the maximum size is 2.9170in/74.091mm.

If the main bearing tunnels are not within tolerance, the chance of bearing failure ('spun' bearings) - the bearings rotating in their housings - is dramatically increased; in fact, to the point that failure is virtually guaranteed if the engine is subjected to any serious stress. When a main bearing spins the oil supply to the relevant big end bearing is cut off, and that bearing will fail immediately. The straightforward checking of the seven main bearing tunnel sizes will prove the integrity of the block beyond doubt.

Tunnels damaged by 'spun' bearings can usually be detected by looking at the bearing tunnel surfaces and checking for circumferential score marks. If an engine has had a main bearing 'spin up,' the tunnel will not be 'on size' and may well be at 'top size' or beyond tolerance. The only solution in this situation is to align hone or align bore (line bore) the bearing tunnels to restore them to standard size. Be

Main bearing tunnel being measured in the second position ...

very suspicious of a block that has any circumferential score marks on any of the tunnel bore surfaces. Minimum factory recommended size main bearing tunnel bore diameters are the safest bet!

The next stage in checking bearing crush is to fit each new pair of bearing shells to the engine block. This checking procedure is designed to make sure that the new bearing shells are being squeezed and held

SPEEDPRO SERIES

... and in the third.

To check the bearing crush, the bearing shells are fitted and both bolts fully torqued. One bolt is then undone (remove the bolt if you prefer), and the gap between the base of the cap and the block is measured with a feeler gauge. Expect the gap to be a minimum of 0.004 inches and a maximum of 0.006 inches.
Notice that the front main cap has been fettled to give more clearance for the oil pump/distributor drive assembly. This allows for removal of the main cap without disturbing the drive.

in the main bearing tunnels to design specification. This check involves fitting each pair of main bearing shells into the block, torquing each pair of cap bolts to the specified torque and then removing one bolt completely and checking the amount of gap that appears between the main cap and the block register.

Caution! - If the gap is 0.002in/0.050mm or less, it is NOT enough. If the main bearing tunnel sizes of the block are on 'bottom size' and the gap is too small, the bearing shells are not the right size (try another set). If the main bearing tunnels of the block are at the 'top size', as measured, and the gap is too small, the main bearing tunnel is too large.

If there's insufficient bearing crush, don't proceed with the engine assembly until the problem has been remedied. It's permissible to remove between 0.001 and 0.0012in/0.0254 and 0.0304mm from the base of the main cap to increase the effective bearing crush. If this doesn't work, the block will have to be align bored or align honed by an engine reconditioner/engine machine shop to restore the correct bearing crush. An alternative, of course, is to find another block which has better sized main bearing tunnels.

With the bearing crush proven beyond doubt to be correct the missing bolt is fitted back in the main cap and torqued to the correct setting. The internal bore size of the fitted bearing shells is then measured with an inside micrometer to see what the actual fitted diameter is. The nominal size (the size you can expect to find) is 2.7530-2.7535in/69.926-69.938mm, while the 'bottom size' (smallest size) is 2.7525in/69.913mm. Top size (the largest size) is 2.7545in/69.964mm. Measure each bearing size in turn, and record the data for future use.

CYLINDER BLOCK

The bolt is then re-fitted and torqued up again.

Caution! - Care must be taken when measuring the internal diameter of the bearing bore as it's very easy to indent the bearing surface with the micrometer anvils and get a false reading. If in doubt about your own ability, engine reconditioning engineers are doing this sort of checking every day and know how to accurately measure bearing bores.

MAIN CAP LOCATION DOWELS

The main caps of XK blocks are located on hollow dowels, and the fit (of the dowels into the block and the main caps onto the dowels) has to be absolutely perfect. The dowels should need a very light tap with a small rubber mallet to fit into the block, and the main caps should require a light tap to fit onto the dowels. If the main caps are not a tight fit, 'fretting' (which is the main caps being able to move around on the block when the engine is running) will ultimately ruin the block.

Fretting is not a very common problem with Jaguar XK blocks, but it does happen occasionally. If the block and main cap matching surfaces aren't marked in any way, then there hasn't been any appreciable main cap movement on the block (though, of course, that doesn't mean that the process of 'fretting' isn't going to start). If you find any loose dowels, you really do need to do something about it before assembling the engine.

To avoid 'fretting', you will need to make custom location dowels which are slightly oversize in relation to the main cap or block (or both). If the fit in the block is correct, for example, and the clearance found to be in the main cap, stepped dowels will have to be made. The dowels may only need to be 0.001-0.003in/0.0254-0.0762mm oversized, which isn't much, and an engine reconditioner or precision engineering works should be able to make new dowels with little trouble.

THRUST WASHER REGISTER - CHECKING

A final, and vital, check that should be

Main bearing bore being measured with an inside micrometer.

27

made before any time and/or money is spent on a block, is that the register (machined recess) that the rearmost thrust washer fits into is still as it was machined by the factory and has no sign of any damage. Although this is not a common problem, the register needs to be checked now and, if any problems are found, the block should be rejected. This surface can be remachined, of course, but it involves some work and expense and normally the block would simply be replaced with another one in better condition.

Chapter 2
Crankshaft & conrods

The crankshaft and connecting rods must be thoroughly cleaned before any work can start. All non-machined surfaces can be wire-brushed and solvents can be used on all surfaces.

Whenever an engine is being rebuilt the connecting rods are generally refurbished first, followed by the crankshaft. Certain parts need to be refurbished before other parts so that the working relationship between the various moving parts is a known quantity before any are machined.

CONNECTING RODS - CHECKING

Several different connecting rods have been fitted to these engines over the years. All early connecting rods were polished on their 'I' beams along the forging flash line and had a drilling up through the centre of the 'I' beam for pressure feed lubrication of the little end. These connecting rods were very well finished for a standard production engine.

Later XK engines have an uprated connecting rod which is not polished along the 'I' beam and is slightly different in design around the bolt head area. There is also no drilling up through the centre of the connecting rod's 'I' beam for pressure feeding the little end. These later connecting rods are the best ones to use.

In all cases, if they're in good condition, the standard XK connecting rods are strong enough for up to 6000rpm operation. The only problem with the older XK connecting rods is that they will most likely have done a lot more work than newer ones.

Because it is likely that you won't know how the engine you have has been treated over the years, it is a good idea to assume that the older the rods the greater the risk of failure. Likewise, the younger the connecting rod and the lower the engine's mileage, the less chance there is of a connecting rod failure.

Caution! - The following sections explain what must be checked on each connecting rod.

Crack testing

Visual inspection might find some cracks or flaws, so do this carefully before submitting rods for further, professional, testing.

All connecting rods must be thoroughly crack tested by an engine reconditioner/engine machine shop using the 'Magnaflux' system to ensure that they are crack free. Experienced operators of this equipment are pretty good at finding cracks.

Additionally, many people have their connecting rods X-rayed in case there is any internal cracking ('Magnaflux' reveals only external cracks). If you can find one that's prepared to do it, X-raying can often be done reasonably inexpensively at a veterinary surgery - though you might raise a few eyebrows at the reception desk ...

SPEEDPRO SERIES

Straightness testing

Most engine reconditioners/engine machine shops have a connecting rod straightness testing jig which allows them to check any connecting rod for correct alignment very quickly. Although XK connecting rods are quite rigid items, it's certainly not impossible to find a bent one. The tolerances on connecting rods are very tight with good reason (inaccuracy causes engine failure). Essentially, the centres of the big end and the little end of the connecting rod must be in perfect alignment. Anything else (more than 0.001in/0.025mm in either plane) is definitely unacceptable. **Caution!** - Fit only perfectly straight connecting rods into any XK engine. Be prepared to find another set of connecting rods if the set you have does not measure up.

With most connecting rod alignment testing machines there is another attachment which allows the piston and connecting rod alignment to be tested once the pistons have been fitted. The connecting rods will have been checked before the pistons are fitted so, if the connecting rod and piston alignment is out, the piston is at fault. Most pistons are accurately made, but there's always a chance that one could be faulty and if undiscovered could cause engine failure. This reasonably simple check will prove beyond doubt the integrity of the rods and their pistons.

Connecting rod bolts

Caution! - Always fit a new set of connecting rod bolts and nuts and check that the bolts take the specified torque without stretching excessively. This means that the overall length of each connecting rod bolt must be measured with a micrometer before the bolt is tensioned.

Hold the connecting rod across the big end in a large vice (which has

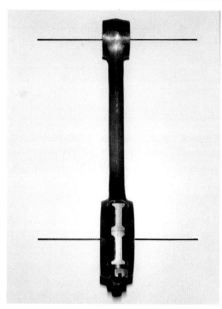

Connecting rod big end and little end bores must be absolutely accurate in these planes.

Connecting rod big end and little end bores must be absolutely accurate in these planes.

jaw protectors fitted) to do the nuts up. Because they're very important surfaces which control the side clearance of the rod, the sides of the big end bearing must not be damaged in any way by the vice.

With the connecting rod nuts tensioned to the correct torque, the average bolt will increase in length by approximately 0.005in/0.127mm (nut torque is nothing more than a convenient way of measuring bolt stretch and, consequently, clamping power!). A vernier calliper can be used to check this stretch, though it will not be as accurate as a micrometer.

Caution! - If the bolt takes the factory specified torque but increases in length by 0.015-0.025in/0.381-0.635mm the bolt is not suitable for use. Although this situation does not occur often, it can and does happen from time to time. If you happen to get one of these bolts and miss this problem, your engine will, most likely, end up ruined. This quick and easy bolt stretch check is very important.

Little end bushings

Check and replace the little end bushing if there is any wear at all. While these bushings do last a along time, they eventually wear out. The little end bushing must be within tolerance when measured, and must be parallel for the whole of its length. Frequently, the bushing will become 'bell-mouthed'. This means that, although the little end bushing might well be on size when measured in the middle, the diameter at the outer edges could well be 0.002in/0.050mm, or more, larger.

The nominal size of the little end bush is 0.875in/22.225mm diameter. Ideally you'll have 0.0002-0.0003in/0.005-0.007mm of clearance between the gudgeon (piston) pin and the little end bushing, which is just enough clearance to allow the assembly of the components. **Caution!** - A clearance of 0.0005in/0.0127mm is

CRANKSHAFT & CONRODS

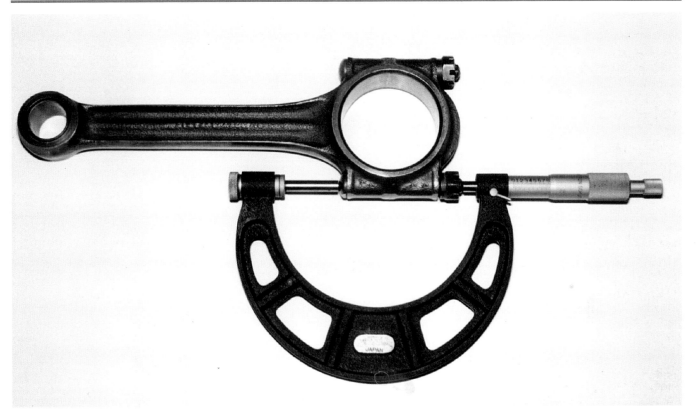

Bolt length is measured using an outside micrometer with the nut untensioned. Then the nuts are tensioned to correct torque and bolt length is measured again.

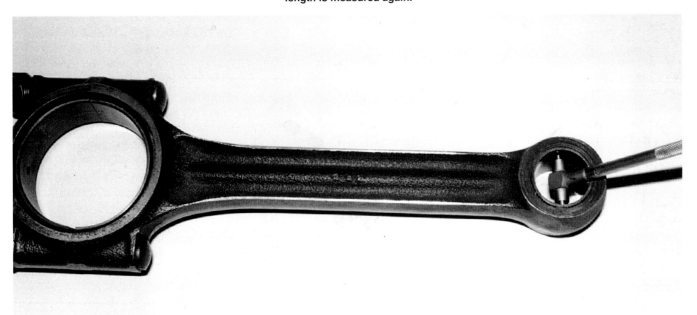

Little end bushings are measured with a telescopic gauge. An outside micrometer is put across the anvils of the telescopic gauge to get the measurement. This is a very accurate way of measuring small bores.

SPEEDPRO SERIES

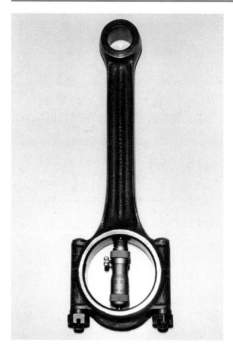
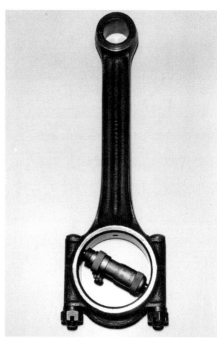
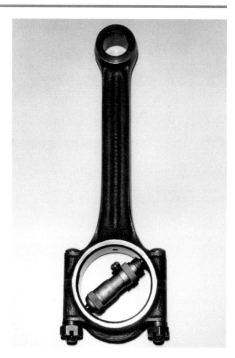

Big end tunnel bore (no bearing shells fitted) being measured with an inside micrometer in the vertical plane, 60 degrees to the left, and 60 degrees to the right.

the absolute limit. The inside diameter of a little end is normally measured using a telescopic gauge (which is then itself measured with an outside micrometer).

Big end bearing tunnel bore

Caution! - The big end bearing tunnel bore must be perfectly round and correctly dimensioned. To check this, the bearing cap is fitted onto the connecting rod, the nuts fitted and then torqued to the specified amount. This is always done with the connecting rod held in a vice (use jaw protectors so that the sides of the connecting rod will not be damaged). The bore diameter of the connecting rod is then measured with an inside micrometer: the specified diameter is 2.2330-2.2335in/56.718-56.730mm. Furthermore, the bore must be exactly round and parallel, so take measurements in several different positions. Having the connecting rod

big end tunnel bore exactly round, parallel, and to the minimum factory specified size (2.2330in/56.718mm) is the ideal.

In many instances engine reconditioners/engine machine shop engineers simply re-size the connecting rods as a matter of course, using the specialised equipment (honing machine) that they have for this purpose. This removes all doubt as to the integrity of the big end bearing tunnel bore. In this procedure, the part-line faces of the conrod and the conrod cap are machined to remove 0.001in/0.025mm of material from each, effectively making the connecting rod tunnel bore smaller by 0.002in/0.050mm (also ensuring that the two surfaces are dead flat). The connecting rod tunnel bore is then re-sized by honing it out to specified 'bottom size.'

Big end bearing crush

The next step is to fit the two bearing

shell inserts into the connecting rod in which they'll eventually be housed when the engine is finally assembled. With the two conrod bolt nuts torqued to the specified amount, undo one of nuts completely and then measure the resulting gap between the cap and the connecting rod with a feeler gauge. Expect the gap to be between 0.004-0.006in/0.101-0.152mm. If the gap is 0.001in/0.025mm, or less, there is not enough 'bearing crush.' If the tunnel diameter is correct at 2.2330 to 2.2335 inches in diameter, the bearing shell inserts are faulty. Try another set of bearing shells.

When the correct bearing crush has been achieved, the nut can be fitted and torqued up again and the inside diameter of the bearing bore measured with an inside micrometer. This is done very carefully (not too much pressure being applied to moving the micrometer) so as not to mark or indent the soft bearing surface

CRANKSHAFT & CONRODS

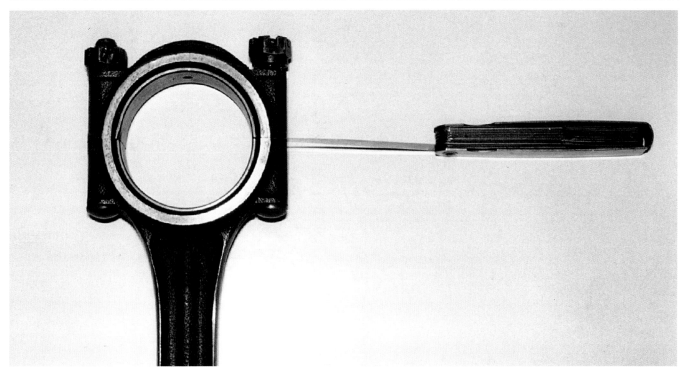

When one bolt is undone the cap must have a gap between it and the connecting rod on the side of the connecting rod cap which has had the bolt undone. The gap is between 0.004 and 0.006 inches.

and thereby gain a false reading (it takes some practice to become adept at doing this). The diameter of each bearing is recorded for future reference when the bearing running clearances are checked. Expect the big end bearing bore diameter (shells fitted) to be from 2.088-2.0885in/53.03-53.047mm.

Although insufficient bearing crush is not a common problem, it does happen from time to time and can ruin an engine (possible spun bearing). It does not take much time to check each connecting rod, it's good engine assembly practise and an expensive engine failure may be averted. Good engine reconditioning/engine machine shop engineers always 'check fit' the components of any engine as they assemble it. The last thing an engine rebuilding workshop wants is an engine to come back damaged, all for the sake of 20 to 30 minutes extra work. They avoid this sort of problem through good preparation and so should you. Take nothing for granted: check everything ...

The goal of all this checking and preparation is to restore the connecting rods as much as is possible to original condition or better. Doing all of this work does not, of course, remove the possibility of metal fatigue and consequential connecting rod failure. After all this work a connecting rod could still fail, the risk of, though, has been reduced to a minimum and is usually quite low.

Caution! - Avoid using connecting rods that you know came from a racing engine that was revved to high rpm. This is a real risk because, although Jaguar connecting rods are of excellent quality, they'll only stand so much stress. Second-hand connecting rods from an engine that has only ever been used at normal road going revs are the safest bet.

CRANKSHAFT
The crankshaft is first thoroughly cleaned and then the journal surfaces are inspected to see what condition they are in. The ideal journal surface has a mirror finish, with no score marks of any description on it. The journals themselves must be perfectly round and 'on size.'

Cleaning oilways
The 'sludge trap' plugs fitted into the throws of the connecting rod must be removed and the contents, accumulated over of thousands of miles, removed. This can sometimes take some work as the plugs can be very firmly located in the crankshaft. Applying heat to the plug by way of

SPEEDPRO SERIES

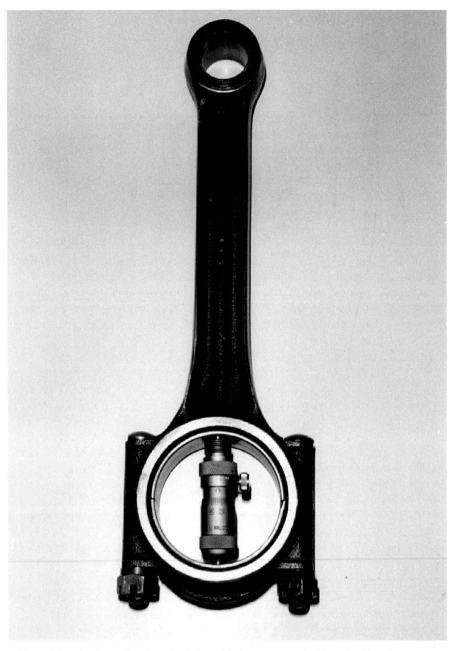

Big end bearing bore (bearing shells fitted) being measured with an inside micrometer. Check the diameter in different places.

an oxy-acetylene torch can sometimes be the only way of loosening the plug (**Caution!** - localised heat should be applied to the plug, not the crankshaft). Buy new plugs and fit them (using a locking compound) when the drillings have been cleaned and after the journals have been reground.

Grinding journals to optimum sizes

With the main bearing shells all fitted into the block, the 'bearing crush' of each combination is checked and the bearing bore size measured and recorded. All seven main bearings could be the same diameter or there could be some small variation, that's within tolerance. At this point the connecting rods will also have all been refurbished, the bearing shells fitted, and the bearing bore diameter size of each connecting rod measured and recorded.

With the actual diameters of the seven main bearings (front to rear order) and the six connecting rod bearings known (front to rear order), the individual journals of the crankshaft can, if required, be reground to suit each bearing bore diameter and to provide optimal running clearances.

There is almost always some slight variation in engine bearing diameters within the factory specified tolerances. If an engine is rebuilt and the crankshaft simply reground to standard journal sizes and the bearings fitted into the engine without checking to see what the clearances are, a very serious risk is being taken, especially if the engine is going to be revved to 6000rpm. The process suggested here removes the possibility of there being tight or loose bearing clearances and reduces the possibility of engine failure.

For maximum reliability the bearing clearances must all be the same and optimal. This may mean grinding individual crankshaft main and big end journals to suit individual bearings. This isn't actually a problem since all journals are individually measured during the grinding process anyway.

The minimum crankshaft to main bearing clearance to have is 0.0025in/ 0.0635mm, with the maximum being 0.003in/0.0762mm. The optimum main bearing to crankshaft clearance is 0.0027-0.0028in/0.0685-0.0711mm.

CRANKSHAFT & CONRODS

The minimum big end journal to bearing clearance is 0.0018in/0.0457mm, while the maximum is 0.0025in/0.0635mm. The optimum big end bearing clearance is 0.002-0.0022in/0.0508-0.0558mm.

Caution! - Standard sized journal fillet radii, as originally ground on all new crankshafts, must be strictly maintained on any reground XK crankshaft. Failure to do so can result in crankshaft breakage (resulting from stress points). All radii must be true radii, as ground, and be polished. Nothing less will do.

The thrust bearing surfaces of the crankshaft will, most likely, have to be reground. The width size can be ascertained by taking the thoroughly clean centre main cap, positioning the new thrust washers onto it on each side, and measuring across the two thrust washers and cap with an outside micrometer. Add 0.003in/0.0118mm to the measured size and this is the minimum width size between the two thrust surfaces of the crankshaft. Absolute care must be taken to ensure that the thrust surfaces (especially the rear facing one) end up smooth, 90 degrees to the crankshaft axis, and have a mirror finish. Optimum crankshaft endfloat at rebuild is 0.003in/0.0118mm. After some running this will become 0.004in/0.1016mm, which is ideal.

The crankshaft must be checked to make sure that it is straight. This means checking to see that the main bearing journals are all in-line. This is done by putting the crankshaft between centres (the centres of a crankshaft grinder usually) and checking each main bearing journal for run out. Ideally, a crankshaft will have no measurable 'run out' whatsoever. All the crankshaft journals must be perfectly round and have a true fillet radius in the corners. There can be no compromises here as failure to regrind a Jaguar crankshaft to correct specifications can lead to breakage.

THRUST WASHERS

Caution! - There are two types of thrust bearing available for Jaguar XK engines: plain white metal ones and heavy duty copper/lead ones. Always fit copper/lead thrust washers (Jaguar original equipment or from a replacement part manufacturer) to the rearmost thrust; the front thrust washer can be a plain white metal one (silver in colour).

If there is a design weakness with the XK Jaguar engine when it comes to high performance use, this is it. The thrust washers (two in total) are 180 degree washers, and not a pair of 360 degree washers (four thrust washers in total) as found on many other engine designs. In normal road use there will be few thrust washer wear problems. However, when the clutch pressure plate is uprated wear problems are likely. This is because the surface area of the thrust washer is quite marginal given the weight of the crankshaft and the axial pressure that the thrust washer has to take.

Caution! - A high pressure clutch cover assembly may well stop the clutch slipping once there is more engine power, but thrust washer wear can become a problem. Replacing the thrust washer on a Jaguar XK engine is major work, so always fit new heavy duty thrust washers whenever they are rebuilt and make sure that the crankshaft endfloat is 0.002-0.003in/0.05-0.076mm (which will wear to 0.003-0.004in/0.076-0.101mm - the minimum specified by the factory). If the endfloat is too great, the crankshaft should be remachined to take oversized thrust washers in order to restore endfloat to minimum specification.

An XK engine that has excessive crankshaft endfloat will have problems which won't be confined to the thrust washers. Too much forwards and backwards movement of the crankshaft when the engine is running results in the crankshaft 'hammering' the thrust washer surface (metal to metal contact) and a reduction in lubrication of the thrust washer surfaces. The oil is not under full engine pressure at this point anyway because thrust washers have grooves in their faces. The excessive fore and aft movement of the crankshaft also gets transmitted to the connecting rods and, if the endfloat reaches 0.010in/0.254mm, can lead to gudgeon (piston) pin circlip failure (the gudgeon pin can 'hammer' the circlips out of the gudgeon pin bore in extreme cases).

Nothing good comes from too much crankshaft endfloat so, while the official factory tolerance is between 0.004 and 0.006in/0.101 and 0.152mm, 0.006in/0.152mm is really too much (unless the engine is definitely going to be dismantled and rebuilt frequently). The reason for this caution is that the thrust washer will wear, so 0.006in/0.152mm endfloat when new will be more like 0.010in/0.254mm, or more, in a fairly short amount of time, especially if an uprated clutch pressure plate is used. It's safe to say that 0.004 inches of clearance is the minimum/maximum endfloat clearance to have on one of these engines. 0.003 inches is quite acceptable with all new parts since, after running-in, what was 0.003 inches will have become 0.004 inches.

Caution! - The crankshaft surface that contacts the rearmost thrust washer must be in perfect condition. That means that the surface must have a mirror finish, like that of a crankshaft journal, it must be absolutely flat and at 90 degrees to the crankshaft axis. It's quite difficult to regrind this area of a crankshaft as the side of the grinding wheel is used. All too often this surface gets 'burned' when it is reground and the surface is not perfect (it becomes mottled, which is just not good enough). When this surface is not absolutely perfect the thrust washers will never last as long as they otherwise would. Inevitably, there is metal to metal contact between the thrust washer and the crankshaft thrust surface which will result in wear and an increase in endfloat. Do not put an engine together that does not have this aspect of its integrity well and truly as it should be.

Caution! - When an uprated pressure plate has been fitted to a Jaguar XK engine, avoid keeping the clutch pedal depressed longer than necessary. This simple precaution will increase the life of the thrust bearings by a considerable margin. Fitting a pressure plate which has enough pressure for the application but not too much more is a well-founded principle. There obviously has to be enough pressure to prevent clutch slip, but some pressure plates are far too strong for anything but competition engines which don't really do many miles before being stripped down and having things like the rear facing thrust washer replaced.

Chapter 3
Cylinder heads

CYLINDER HEAD CHOICE

There are effectively three types of cylinder head readily available for these engines –

The first is the B-type cylinder head, as found on all Mark II and S-Type Jaguars, for example, which has curved inlet ports offset $3/4$in from the inlet valve.

The second is the straight port cylinder head as found on Mark X Jaguars, E-Types and the XJ6. The ports of the straight port cylinder heads are, in actual fact, offset by $5/16$ in from the centreline of the inlet valve, which means that the inlet port is not dead straight, just straighter.

The third is the larger inlet valve straight port head which is found on the Series III XJ6 engines. The Series III XJ6 cylinder head is the most desirable head of the three because of the larger inlet valve head size and the slightly larger inlet ports. It's fair to say that the Series III cylinder head offers the greatest improvement in engine performance (up to certain point) for the least amount of reworking. This makes it the ideal cylinder head for those on a budget. This cylinder head can be fitted to any 3.4 or 3.8-litre engine, provided the two rearmost waterways are tapped and plugged.

What has to be remembered here is that when Jaguar built and sold these cars new they had to deliver reasonably good miles to the gallon figures. Jaguar achieved this, to a certain extent, by restricting the inlet port flow. They simply could not make these engines with high flowing cylinder heads, more suited to a racing car, and still have them delivering good fuel economy. The upshot of this factor is that none of these standard production, factory-made cylinder heads are right for a performance orientated engine in their standard condition. The inlet ports must be reworked if there's to be a substantial improvement in engine power.

The performance potential of all three cylinder heads (once modified with enlarged inlet ports) is very similar. The fact is that any of these cylinder heads - correctly ported for the particular application - are very suitable for any of these engines (a point which is often lost).

Having a cylinder head that flows a huge amount of air on a flowbench does not necessarily mean that that particular cylinder head is going to be 100% right when fitted on to an engine. It might well be brilliant at maximum rpm but less efficient at lower rpm compared to a smaller ported head. The smaller ported head on that very same engine could well deliver much more usable power and torque. The true end use of the engine needs to be taken into consideration before any cylinder head is modified.

Another factor here (one that usually leads anyone modifying a Jaguar XK engine to a straight port cylinder head), is the fact that if sidedraught Weber or Dellorto

SPEEDPRO SERIES

When fitting an XJ6 cylinder head onto a 3.4 or 3.8 block plug these two waterway holes.

carburettors are going to be used, appropriate inlet manifolds are more readily available for a straight port head. Given the choice, everyone seems to want a straight port cylinder head, and the fact that the B-type cylinder head can be reworked to match the straight port head doesn't usually come into the equation.

The valve head sizes for production cylinder heads (up until the XJ6 Series III engine) were all the same: 1.750in diameter for the inlets and 1.625in for the exhaust valves. The XJ6 Series III cylinder head, however, has 1.875in diameter inlet valves and the usual 1.625in exhaust valves. The inlet valve seat insert is also larger in the Series III cylinder head.

Caution! - Before any XK cylinder head has any time and money spent on it, the camshafts should be checked for free turning by hand. The purpose of this is to check for a warped cylinder head. In some instances, cylinder heads have been excessively overheated and are well and truly warped. The gasket face might well have been trued up by being refaced, but the camshaft tunnels most certainly won't have been re-align bored. XK cylinder heads have been known to warp more than 0.080in/2mm, and this means that when each camshaft is fitted into the cylinder head (with the valves removed) the camshafts cannot be turned freely by hand. The engines will run like this (unbelievable as it may sound) and the camshafts might not break (at least not straight away) but the situation is far from ideal and must be avoided.

If the cylinder head is basically 'straight', you should be able to turn the camshafts by hand. This very simple test sorts out a 'good' cylinder head from a bad one.

It is not recommended that the camshaft tunnels in a cylinder head be align-bored to correct for warpage. Other considerations include the fact that the tappet bores will no longer be at absolutely 90 degrees to the camshaft axis. Find another cylinder head in good condition.

EXHAUST PORTS (ALL HEAD TYPES) - MODIFICATION PROCEDURE

The exhaust ports of all of these production XK cylinder heads are the same and all need a minimum of reworking. The ports are really

CYLINDER HEADS

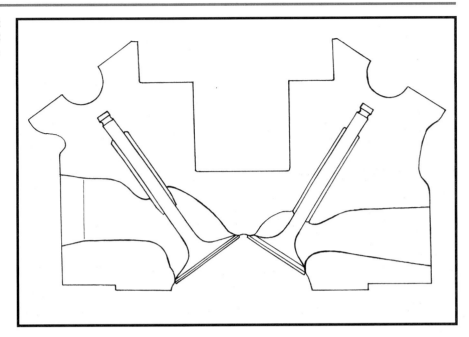

Series III cylinder head: standard inlet port and valve (left) and standard exhaust port and valve (right).

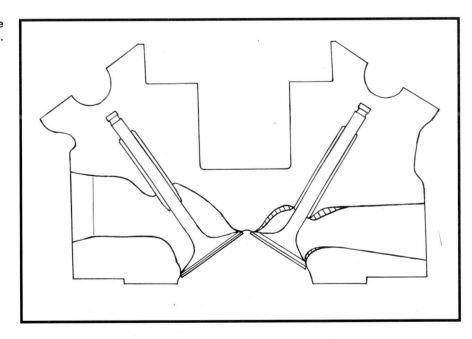

To modify the exhaust ports, remove the material in the shaded areas.

quite large just as they come and should only really need to be decoked (decarbonated). The valve throats and the turns into the ports, on the other hand, do need to be reworked (i.e. have metal removed).

The valve throat as standard is angled and 1.375in/34.92mm in diameter adjacent to the aluminium of the cylinder head casting and 1.400in/35.56mm adjacent to the inner valve seat edge. This, of course, is essentially the inside diameter of the valve seat insert. This diameter can be opened out to 1.450in/36.83mm (parallel bored). The valve seat insert can be ground out using a tungsten carbide rotary file and finished with a grinding stone. Either of these two cutting tools can be 'spun up' in a high speed drill. The tungsten carbide rotary

39

SPEEDPRO SERIES

file will work well at a relatively slow speed (1200-1500rpm). The mounted point grinding stone, on the other hand, needs to be turned at a higher speed than this, otherwise it will wear away quite quickly. Some pistol drills do spin at 3300rpm (some even more) and this is often enough to get the job done without going through too many grinding stones. The diameter of the grinding stones needs to be about 1in in diameter.

Use inside calipers, set to 1.450in/ 36.83mm, to monitor progress and measure the final size of the ground out valve throat. With the valve seat's inside diameter opened out to the specified size, the aluminium of the cylinder head is blended (smoothed) back into the port. The valve guide boss is also reduced in size.

The aluminium of the standard cylinder head casting usually overlaps the valve seat insert and, consequently, reduces the actual diameter of the bore of the insert. It does not take too much work to remove all the excess aluminium and end up with what is essentially a parallel bored hole with no roughness adjacent to the valve seat. The short side of the turn from the valve throat into the exhaust port proper, can be re-radiused so that the radius is as large as it is possible to have it. The removal of the aluminium is done using a rotary file or burr as opposed to a grinding stone (aluminium from the cylinder head will clog the grinding stone).

The exhaust valve seats, as ground in the cylinder head valve seat inserts, are at least 0.100in/ 2.54mm wide. The shape and sizing of the standard valve seat insert is excellent for standard road use, but improvement is possible for performance orientated engines. The standard valve seats are very hard wearing.

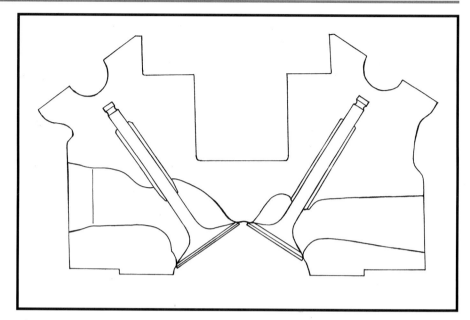

A completed exhaust port (right). Reworking is confined to the valve throat and the turn into the actual exhaust port. The valve throat has been ground out to 1.450in in diameter, the guide boss has been reduced in size and the port turn contour reshaped (re-radiused).

Roughly worked exhaust port. This view looks into the exhaust valve throat/exhaust port from the combustion chamber.

The valve seat in the cylinder head is narrowed down by an 'inner cut.' This means that the outer diameter of the exhaust valve seat must remain at the standard 1.625in/ 41.27mm (valve head diameter size).

CYLINDER HEADS

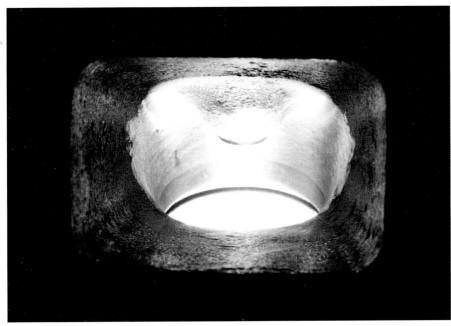

A completed exhaust port viewed from the exhaust manifold side of the cylinder head.

are 'K-Lined' to restore the valve stem to valve guide running clearance. This is a proprietary process which many engine re-conditioners or engine machine shops use, which involves fitting an insert into the original valve guide to restore correct clearances.

INLET PORT (ALL HEAD TYPES) - MODIFICATION CHOICES

With the inlet ports, the idea is to make them only as large (with a smooth contour, so that the engine develops maximum power) as is absolutely necessary. Having the inlet ports as large as it is physically possible to make them, is not the way to best all round engine efficiency. For an XK engine which is not turning more than 6000rpm, there is no advantage in

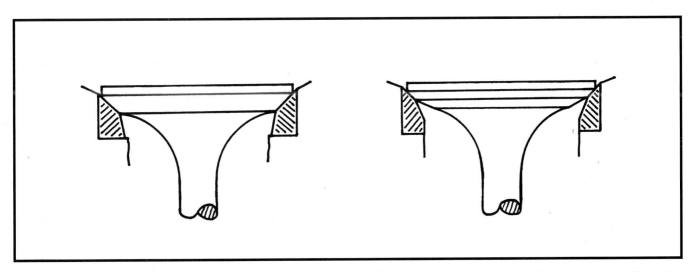

This diagram shows the standard exhaust valve seating (left) and a modified valve and valve seat (right). The standard valve throat is 1.375in in diameter at its narrowest, whilst the modified one is 1.450in in diameter and parallel. Valve seat contact width is 0.080in on the modified cylinder head as opposed to the usual 0.100 to 0.110in.

With the valve seat insert parallel bored to 1.450in/36.83mm in diameter, a small 60 degree inner cut (chamfer) is used to increase the inner diameter of the exhaust valve seat diameter and, as a consequence, reduce the exhaust valve seat width to 0.080in/2.03mm.

Exhaust valve guide bores do wear over time, and the ideal valve stem to valve guide bore clearance is lost. It's recommended that the exhaust valve guides are not removed for replacement of worn valve guide bores (unless damaged) but, instead,

having very large, 'out to the water jacket' sized inlet ports.

There are essentially two ways of porting the inlets on a B-type or straight port XK cylinder head. The first is to leave the actual inlet port diameter essentially standard, enlarging instead

41

the area around and at the valve guide to improve airflow. The amount of reworking required will take someone experienced with porting cylinder heads who has the right equipment about 10-12 hours. The improvement in engine performance for this amount of reworking is very worthwhile and the work is hardly major since the inlet valve guides are not removed. The actual inlet ports (the first 2.5in/63.5mm into the port, from the side of the cylinder up to the valve guide) are, essentially, left as standard at 1.375in/34.92mm in diameter.

The decision over whether or not to replace the inlet valve guides on any XK cylinder head essentially revolves around whether the engine is going to be for road or pure racing use. Retaining the standard inlet port diameter (1.375in/34.92mm) means you can leave the valve guides in the cylinder head.

Increasing the inlet port diameter from standard on a B-type cylinder head or a straight port cylinder head means that the valve guides really do need to be removed. It's also fair to say that a better job can be made of the inlet ports if you don't have to work around the guides. Having said this, because of the relatively small amount of material that has to be removed from the inlet ports, it's generally regarded as expedient to work around the guides.

Any cylinder head that will, essentially, retain the standard sized inlet ports can be successfully ported while working around the protruding valve guide. It does take time to do this correctly and carefully, and it's a fiddly task, but it can be done and does save the extra work involved in removing the old guides and, subsequently, installing new ones.

A further consideration here is the fact that original cast iron valve guides are not as wear resistant as original guides which have had 'K-Line' inserts fitted. K-Line inserts are very wear resistant and allow very tight valve to guide clearances to be used with little risk of seizure. This means that the worn valve guides don't need to be removed from the cylinder head, unless they're damaged. By this proprietary method the original stem to guide fit can be restored perfectly without removing and then replacing the valve guides.

Many engine reconditioners or engine machine shops fit K-Line valve guide inserts, and they are good value on the basis of lasting ability and effect. Locating an engine reconditioner which can install these valve guide replacements isn't difficult these days. K-Line inserts can also be replaced as many times as is necessary, but expect them to last 2-3 times as long as standard cast iron valve guides in this application.

In recent years, tetraethyl leaded fuel has been phased out in most developed countries around the world. Depending on the unleaded fuel available, exhaust valve sticking problems can occur. The solution to this is to K-Line the valve guides (all exhaust guides as a minimum). The high copper content of the guides keeps the valve stems/valve guides clean enough to prevent valve sticking. This is another very good reason for K-Lining both the inlet and the exhaust valves.

Large inlet ports

The second approach to porting the inlets is to leave no stone unturned and do everything possible to improve them and to maximise flow potential. This means removing the inlet valve guides from the cylinder head, and opening up the inlet port diameters to 1.5in/38.1mm in diameter. This task amounts to major reworking of the cylinder head and will take about 30-35 hours, or more, to do. The advantages of carrying out this amount of work is better top end power, with the engine having what is best described as a more urgent rate of acceleration under full throttle from the mid range (4000rpm) through to 6000rpm.

Be aware that road going fuel economy with a large port cylinder head (1.5in/38.1mm) will be noticeably reduced compared to an essentially standard inlet port diameter cylinder head. Unfortunately, once the inlet ports are opened up the economy is gone for good. Be sure you are confident about the true end use of the engine is before you start drastically altering the cylinder head!

VALVE GUIDES (ALL HEADS) - REMOVAL & REPLACEMENT

When removing the valve guides from any aluminium cylinder head, the usual recommendation is to heat the cylinder head up and 'drift punch' the old guides out of the head. This is all very well, but it's not the best way of removing valve guides since there is the possibility of damage to the original valve guide drilling in the cylinder head. The valve guides will almost always have been in the cylinder head for many years and, often, when the old guides are removed, the actual drilling in the cylinder head gets damaged so that it is no longer the size that it was originally bored to by Jaguar. When a new standard size replacement valve guide is fitted into a damaged drilling the interference fit is no longer as it should be. This means the valve guide could end going up and down in the cylinder head instead of the valve stem going up and down in the valve guide bore. Serious trouble is on the way if this is happens ...

CYLINDER HEADS

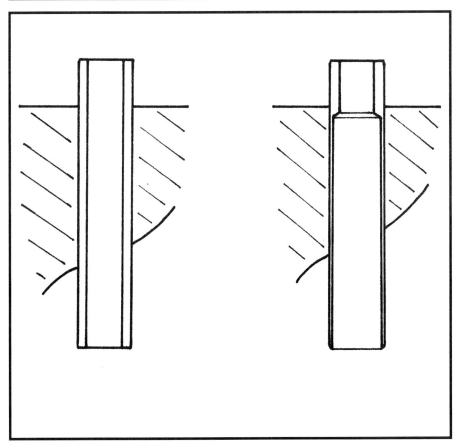

The valve guide on the left is a standard one. The valve guide on the right has been drilled out from the combustion chamber side of the cylinder head. Note that it has not been drilled right through.

To avoid this situation, the cylinder head needs to be mounted on to a milling machine, or a 'Kwikway' head shop machine, and the guides bored out from the combustion chamber side of the cylinder head. This means accurately locating the centre of the valve guide and removing most of the valve guide material with a drill, until the shell of the guide has no material strength and becomes loose in the cylinder head. The drill size to use is 7/16in diameter. The drill is not taken right through the valve guide but, instead, is stopped short of the end of the valve guide by 3/8in (9.5mm).

The guide that protrudes into the inlet port is also thoroughly cleaned with a rotary wire brush to remove all traces of carbon so that only the guide is pushed through the guide bore.

If, after the guides have been drilled as described, they still refuse to budge, the cylinder head should be heated to 230F/110C and a 3/8in (9.525mm) diameter drift punch (a piece of mild steel rod will do) used to move the valve guides out of the cylinder head (they should tap out).

If the camshaft end of the drilled out valve guide breaks off, another drift punch, 0.495in/12.57mm in diameter, will have to be made (turned down from suitable diameter mild steel bar) with a 1in/25.4mm long pilot section which is just under 7/16in (11.11mm) in diameter (this must just fit into the drilled hole in the valve guide). This time the old valve guide is drifted out from the camshaft side of the head. With the outer diameter of the drift punch 0.005in smaller in diameter than the drilling in the cylinder head, and the pilot section of the drift a close fit into the valve guide, there is no possibility of damaging the aluminium of the cylinder head.

Sometimes, the old valve guide will break up, leaving debris within the aluminium of the cylinder head drilling. The critical factor here is to avoid damaging the factory bore valve guide holes in the cylinder head at all costs, even if it does take a bit of extra work. Do whatever it takes to make sure the original valve guide drillings are kept 'on size'.

If you want extra durability you can have new replacement valve guides K-Lined, even though the new cast iron guides will last for quite a long time from new.

Once the inlet ports have been opened out, new guides can be fitted to the cylinder head in the recommended manner, i.e. freezing the replacement valve guides and heating the cylinder head. The new valve guides then literally fall in and, once the temperatures have equalised, will be held firmly in place.

The diameters of the replacement valve guides need to be measured with a micrometer to make sure that they are on 'top size'. The bores of the valve guide holes in the cylinder head also need to be measured, with a telescopic gauge, to make sure they, too, are 'on size.' This is to ensure that the correct interference fit will be present when the guides are installed. Optimum interference fit is 0.002-0.0025in/0.0508-0.0635mm. **Caution!** - 0.0015in/0.0381mm is not enough, the fit is too loose.

SPEEDPRO SERIES

If a valve guide drilling has become oversized, and the standard recommended amount of interference fit is not possible, a valve guide will have to be made that does give the recommended interference fit. This is achieved by turning down another type of valve guide (with a larger outside diameter) to the correct size to restore the interference fit. **Caution!** - Do not be tempted to fit a valve guide that does not have enough interference fit. The prospect of having the valve guide going up and down in the cylinder head, instead of the valve stem going up and down in the valve guide, is a very real one and could lead to an expensive engine failure.

INLET PORT (B-TYPE HEAD) - MODIFICATION PROCEDURE

The B-type cylinder head inlet port is offset from the centreline of the valve by approximately 0.75in/19.05mm. The port actually has quite a smooth contour and, although it's not a poor flowing design, it can be improved upon for a high performance application. The B-type cylinder head is generally underated by and is almost always viewed as being inferior. This is not the case, and sells the B-type cylinder short. The B-type cylinder head is a very good cylinder head and can run with the best of them after suitable reworking.

The standard B-type cylinder head has an inlet port opening diameter, at the side of the cylinder head, of 1.625in/41.27mm. The port tapers down to approximately 1.312in/33.33mm in diameter after the first 1in (25mm) or so, making it the smallest port diameter used on XK cylinder heads. This diameter is more or less maintained (it gets a bit larger if anything) through to the valve guide, but it is not all that uniform. B-type heads do need to have material removed from the inlet port to increase the port size from 1.312in/33.33mm to 1.375in/34.92mm. This is more of a cleaning up process rather than a major metal removal process. Careful reshaping of the inlet ports can really

Left: B-Type inlet port modified to a nominal 1.5in diameter (valve guide has been removed). Right: Standard B-Type inlet port, nominal diameter is 1.312in.

CYLINDER HEADS

pay dividends with the B-type cylinder heads.

There are essentially two options when porting a B-type cylinder head –

The first option involves enlarging the inlet ports slightly from 1.312in/33.33mm to 1.375in/34.92mm in diameter. You can leave the inlet guides in place and worked around, or replace them with new items: working around the valve guides is usual. This inlet port sizing is ideal for road going engines and other applications where maximum low down to mid range torque is essential. This inlet port sizing is not inferior to the following, larger, port option; it's just that the application should decide the option for you.

The second option involves the removal of the inlet valve guides and the opening out the inlet ports to approximately 1.5in/38.1mm in diameter. This is the ideal choice for a racing engine which will not be expected to perform miracles under 3000rpm.

The easiest to check that sufficient material has been removed from each inlet port is to check the size of the port with a gauge. Because XK inlet ports are round, an old inlet or exhaust valve ground to the right size is ideal for doing this. Another alternative is to use a washer of the right outside diameter fitted to a length of stud (threaded rod). The washer is 'trapped' on the stud or bolt by having a nut fixed on each side.

The diameter of the washer or valve head should be smaller than the required diameter of the inlet port by 0.025in/0.635mm. This is to make sure that the inlet port is not made too large. The gauge goes into the port as a 'rattling good fit' (this means

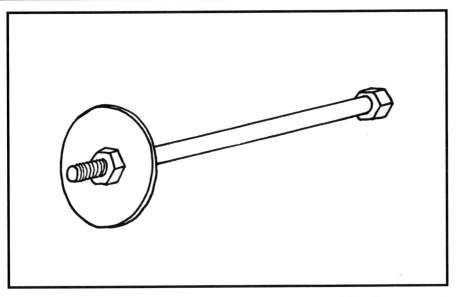

A gauge such as this can be used to check whether sufficient material has been removed from the inlet port. The outer diameter of the washer can be ground down to the correct size.

Material is removed from the roof (A) and the side wall of the inlet port (B). The floor of the port (C) is essentially left untouched, as is the other side wall (D).

SPEEDPRO SERIES

Views of an inlet port from the combustion chamber. The majority of the material that is removed from the inlet port comes off around the valve guide (side wall and roof). This inlet port was modified with the valve guide in place.

CYLINDER HEADS

the gauge will pass through the port without becoming jammed, but that there will be a minimum of clearance around it). By making the gauge undersize, and fitting it as described into the inlet port, the port will finish up more or less 'on size.' Make the gauge 1.350in/34.29mm in diameter for an option one type cylinder head. The gauge must be able to go into the inlet port right up to the valve guide (if the valve guide is in place), and past the valve guide location if the valve guides have been removed from the cylinder head.

Material removed from the inlet port should not be taken from the 'floor' of the port, but mainly removed from the 'roof' and one side wall of the port. The 'roof' of the inlet port is nearest the camshafts, and the side of the port from which to remove metal is the side that the valve guide protrudes from. This method of material removal maintains the standard height of the inlet port in the cylinder head and keeps the short turn into the valve throat as large as it is practicable. This means that the centre of the floor of the port will remain, basically, untouched.

The B-type cylinder head inlet port curves, with the valve guide protruding from the short side wall of the turn. When the inlet port is reworked, the material removed from the side wall is taken from the wall from which the valve guide protrudes. By reworking the inlet port in this way it is, to a degree, straightened. Up to 0.0625in/1.58mm is removed from the roof and one side wall of the inlet port. The fact that the port might end up a little more than 1.375in/34.92mm in diameter is not cause for concern, but you should try to maintain the correct diameter up to the valve guide or edge of the valve guide location (if it's been removed). At this point the cross sectional area of the inlet port can be increased without detriment to airflow. The port shape is changed at this point as the floor of the port is flattened.

Note: inlet ports are right and left handed in the cylinder head but this is of no consequence as the ports are, apart from this factor, identical.

The B-type cylinder head, unlike the straight port cylinder head, does not have a valve guide boss (the valve guide protrudes directly out of the inlet port wall). After the valve guide, the port flares out to meet the inside diameter of the valve seat insert, which is 1.5in/38.1mm in diameter at this juncture. The port also turns as it goes into the valve throat area (a complex shape). It is not necessary to enlarge this part of the inlet tract by too much, though some reworking is necessary to smooth the contour of the turn from the actual port into the valve throat.

The 'valve throat' of the inlet tract is that area which comprises the valve seat and a portion of the turn into the port (a distance into the inlet port from the valve seat of 0.5in/12.7mm on the short side of the turn (contacts the floor of the port) and 0.75in/19.05mm on the long side of the turn into the port (contacts the roof of the port).

The floor of the inlet port on the valve guide side is also flattened off slightly, which reduces the radius of the turn on this side of the inlet port, and also increases the cross sectional area of the port in this area. This process has at times been termed 'flat flooring,' and is applied to all XK cylinder heads.

The inlet valve seats are usually 0.100in/2.54mm wide, or more, after a regrind or two. Most engine reconditioners/engine machine shops

The valve guide was removed before this port was worked on. It's much easier to modify an inlet port with the valve guide removed. This port has been roughed out using a rotary file.

47

SPEEDPRO SERIES

The original valve guide was removed before work commenced. The port has been roughed out with a rotary file and a new valve guide fitted. The valve seat is still standard.

simply regrind the valve seat to clean it up and make no attempt to narrow the valve seat cut into the cylinder head. The valve seat (as cut by the factory) is not all that good in absolute terms, but is good enough for a standard production engine. For high performance applications, the valve seats need to be reground so that they are 0.060in/1.52mm wide. The inlet valve seat width could be slightly less (ideally 0.050in/1.27mm, but never less). The outer diameter of each valve seat, as ground into the cylinder head, needs to be the same size as the inlet valve diameter which is 1.750in (44.45mm).

The inlet valve seats on just about all XK cylinder heads are just about proud of the aluminium of the cylinder head by around 0.010-0.020in/0.254-0.508mm. The actual valve seat contact area should be re-cut so that the outer diameter is the same as the inlet valve size (1.75in/44.45mm in the case of the B-type cylinder head). The 45 degree valve seat is re-cut, and a radius used to blend the 45 degree valve seat into the valve throat on the port side of the valve seat. **Caution!** - This radius can be cut by hand, but care is needed - one slip and the valve seat will have to be re-cut!

Large inlet port
For motor racing purposes, the B-type cylinder heads inlet ports can be opened out more radically so that they become so-called 'large port' cylinder heads. The inlet valve guides are almost always removed to do this, and the ports are then opened out to a maximum size of 1.5in/38.1mm in diameter. This target diameter involves the removal of between 0.125in/3.175mm and as much as 0.187in/4.76mm of material from the inlet port. It is possible to enlarge the inlet ports of a B-type cylinder head to this extent with the valve guides in place and make a very satisfactory job

CYLINDER HEADS

The inlet valve seat insert has a reasonably large valve seat contact area (note how the insert is slightly proud of the chamber face).

of the work. Working around the guide is of course the problem, a job which requires care and patience should you choose to adopt this method.

For a big port head the gauge is sized at 1.450in/11.43mm diameter. Material will have to be removed from the floor of the port, but this is minimal. After a distance of 3in/76mm in from the inlet manifold face, very little material is removed from the floor. The majority of material is taken off the roof of the port and the side wall (the one from which the valve guide protrudes).

If all of the material is taken off this side wall, the casting will be getting a bit thin in one or two places and there is some slight risk of breaking through into the water jacket. Some material, therefore, can be taken off the opposite wall. There is usually plenty of material around the inlet ports on B-type cylinder heads. Removing the material carefully with the aid of a gauge is the easiest way to start taking the metal out of the port and establishing a general port contour. The gauge diameter size is then increased by 0.025in up to 1.475in in diameter. This is done to give a safety margin during porting to avoid going too large.

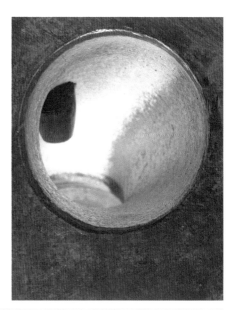

Enlarged inlet port on a B-Type cylinder head. The valve guide has been removed.

49

SPEEDPRO SERIES

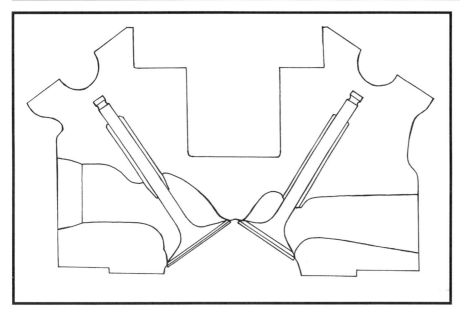

Standard Series III inlet port showing the inlet valve guide and, more importantly, the valve guide boss clearly intruding into the air stream (valve and port on the left).

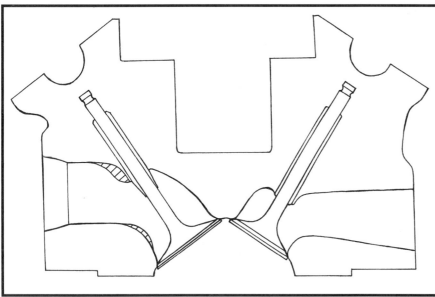

Remove the shaded areas of the inlet port.

The B-type cylinder heads inlet ports are more difficult to enlarge than the straight port cylinder heads because of the curvature of the inlet ports. The shape is more complex.

INLET PORTS (STRAIGHT PORT HEADS) - MODIFICATION PROCEDURE

The standard straight port cylinder head has an inlet port opening diameter of approximately 1.625in/41.27mm (all XK heads are the same here). Within an inch of going into the cylinder head from the side of the head, however, the inlet port becomes approximately 1.375in/34.92mm in diameter which it essentially maintains through to just before the valve guide boss. The inlet port itself, from the side of the cylinder head to the start of the valve guide boss, is not particularly undersized, and is, in fact, quite suitable for all road use and many racing applications. The standard Series III inlet port is slightly larger in diameter than both the standard B-type and the earlier straight port cylinder head.

On the straight port Series III

CYLINDER HEADS

Completed inlet port with the valve guide boss still in place but reduced in size as much as possible.

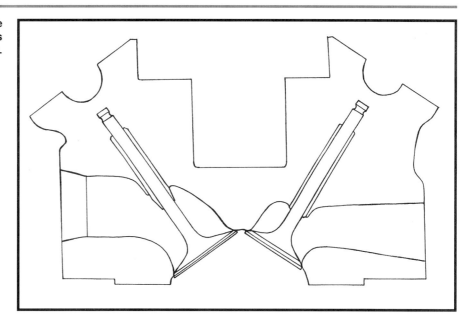

The valve guide was removed before the inlet port was re-worked and then re-fitted

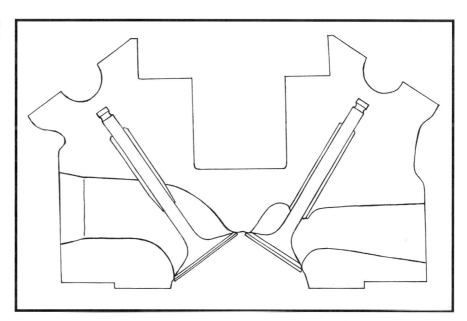

cylinder head, the valve guide boss and the way in which it is blended into the actual port configuration is the main problem. This part of the inlet port is what has to be altered to make a substantial change to the inlet flow. All other parts of the inlet tract can be left as standard.

Improving the flow through the inlet port involves removing aluminium from the inlet port wall adjacent to the valve guide on the short side of the slight turn that is present (increase the cross sectional area of the port at this point). The actual valve guide boss is reduced in size to the minimum possible by carefully working around the valve guide.

When looking into an inlet port, the valve guide and the guide boss are clearly an obstruction which, although it can't be removed, can at least be reduced in size. The cross sectional area of the port is reduced substantially in this area because of the shape of the valve guide boss.

It's possible to taper turn the end

51

SPEEDPRO SERIES

A completed inlet port. The valve guide was removed to do this port. This view is from the inlet manifold side of the cylinder head.

of the valve guide that protrudes into the port a little bit before the new valve guides are fitted into the cylinder head. However, do not turn the valve guide to a knife edge as this will reduce the material strength of the valve guide and bits might break off and go into the engine. In the interests of maximum reliability, it is recommended that the replacement valve guides not be altered in any way.

If the valve guides have been removed, the guide bosses can be removed from the inlet ports altogether, resulting in exceedingly neat looking inlet ports.

Large inlet port
If the engine is being prepared for racing purposes, or you want to have a racing engine which gets occasional road use, it will prove beneficial to remove the valve guides and enlarge the inlet ports to 1.5in/38.1mm in diameter and, in so doing, remove the valve guide bosses completely. Consider this to be the ultimate measure required to extract the maximum possible engine efficiency at high rpm. There will be some loss low down in the rpm range, however, due to the enlarged port diameter. A racing engine is rarely going to be operating below 3500rpm and the loss of engine torque at lower revs will not be noticed.

The diameter of the inlet port is enlarged by 0.125in/3.17mm. This means making the port 1.5in/38.1mm in diameter. Although there is sufficient material in the inlet port to do this, there can be problems that are not necessarily apparent when starting work. In many instances, the engine that donated the cylinder head will have been used for many years without a rust inhibitor in the water. This could mean that the cylinder head has a fair bit of corrosion in the water jacket (where you can't see it). The end result of this could be that when removing material from the inlet port the wall thickness will not be as it was when originally cast by the factory. It then becomes quite possible to go through the water jacket of the inlet port.

CYLINDER HEADS

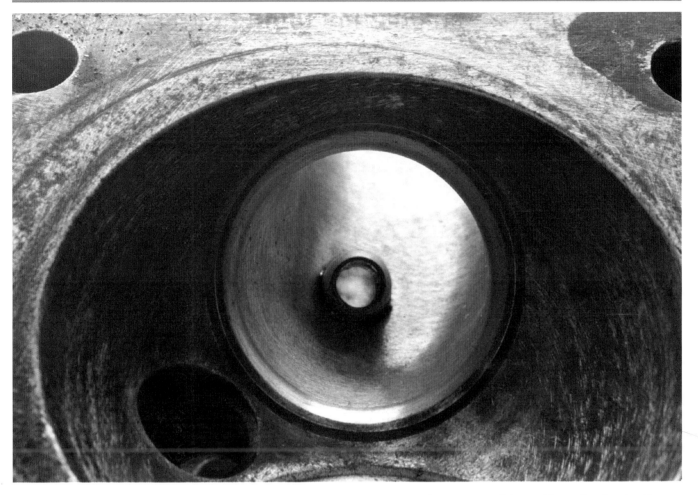

This inlet port was reworked with the valve guide removed. The valve seat has been narrowed down with a radius forming the transition from the valve throat to the inner edge of the valve seat.

The telltale sign of a badly corroded cylinder head is a lot of corrosion around the waterway slots on the head gasket face. The other thing that comes into this is the fact that there can have been core shifts, or some porosity, in the casting when the cylinder head was made, meaning that there is less solid wall in the inlet port than you might reasonably expect there to be. These problems only show up when quite a lot of material is removed from the ports. In standard condition, these cylinder heads are almost always sound. If you do grind into the water jacket, the cylinder head will have to be 'tig' welded to repair it, or another cylinder head found and work started all over again.

Caution! - In most instances you'll be able to successfully modify the cylinder head to large port dimensions, but you do need to be aware of the potential problems. There is definitely some risk of breakthrough attached, and be aware that, ideally, the valve guides should be removed so that the valve guide bosses can be ground away.

It is possible to enlarge the inlet port in the vicinity of the valve guide boss by 0.125in/3.17mm with the guide in place, but it does take a lot of extra time to do this. If the work is carried out correctly, the result is just as efficient as when the valve guides are removed. The problem, however, is that it can get a bit wearing working in the confined space available and corners are likely to be cut. This might well result in a cylinder head that isn't as good as it could, or should, be. The removal of the valve guides, therefore, removes the general possibility of error through this factor.

The port diameter varies as you go on into the port. The diameter at the side of the cylinder head

SPEEDPRO SERIES

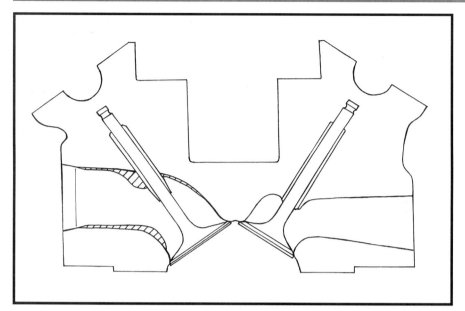

The inlet port on the left needs material removed from the shaded areas. Remove as little material as possible from the port floor and take the bulk from the roof.

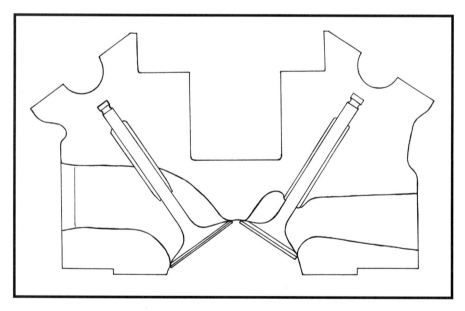

Completed inlet port on the left, and completed exhaust port on the right.

(where the inlet manifold bolts on) is maintained at 1.625in/41.27mm. However, where the port originally tapered down to 1.375in/34.92mm, the diameter will now be 1.5in/38.1mm. This is an increase in port size of approximately 15%. The diameter of the port at 1.475-1.50in/37.46-38.10mm is maintained through to just a 0.5in/12.7mm before the edge of the valve guide location. At this point the diameter is slowly increased to 1.54in/39.11mm.

VALVE SEATS (ALL HEADS) - WIDTHS

The valve seats of most XK cylinder heads will have been reground a few times in the life of a standard engine. Therefore, in most instances, the valve seat in the cylinder head will no longer be to standard original specifications, and will certainly not be suitable for a racing engine.

Frequently, the actual valve seat contact area of the inlet valves will end up at between 0.090in/2.28mm, or

CYLINDER HEADS

This 1.75in inlet-valved straight port cylinder head shows the totally standard port on the right and the fully modified inlet port on the left.

more, which is too wide. The finished size of an inlet valve seat needs to be just 0.060in/1.52mm.

An exhaust valve seat contact area that has been reground a few times will almost always end up approximately around 0.100in/2.54mm wide, or more. This should be reduced to 0.080in/2.03mm.

The actual valve seats (for inlet and exhaust valves) as ground into the cylinder head are almost always on size on the basis of outer diameter. This means that the valve seat insert width is reduced on both inlet and exhaust valve seat inserts by taking material away from the valve seat throat side,

using a 60 degree inner cut or, better still, radiusing the inner diameter of the valve seat insert. Engine reconditioners/engine machine shops will usually use a 60 degree inner cut. Radiused valve seats are better, but more difficult to do. **Caution!** - Valve seats can be radiused by hand very successfully, but extreme care is required as one slip will damage the valve seat and it will have to be re-cut. The exhaust and the inlet valve seats are treated in a similar manner.

VALVES & SEATS (ALL HEADS)

On the basis of strength and quality of

materials, the standard XK valves are quite suitable for all high performance and racing applications using up to 6000rpm.

Caution! - Always fit new exhaust valves to any engine that is being rebuilt with high performance in mind. This is because of the hard life that exhaust valves have because of extreme heat. The valve heads of old, well used, exhaust valves have been known to fall off with much engine damage resulting!

Inlet valves, on the other hand, do not get such a hard time in an engine and, provided they are in good overall condition, can be re-used after

SPEEDPRO SERIES

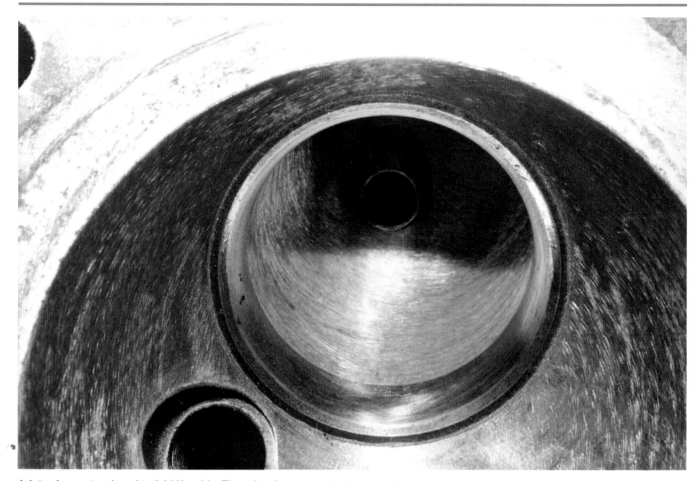

Inlet valve seat reduced to 0.060in wide. The valve throat area has been radiused using a hand held die grinder/high speed pistol drill with a ³⁄₈in diameter by 1in long grinding stone: this method takes some dexterity but is entirely possible.

they have been reground. However, given the cost of these items, it is difficult to justify using old inlet valves in a high performance application. Jaguar engine parts are not particularly expensive. As a general rule, if the valve stems show any measurable wear (measured with a micrometer) the valves should be replaced.

The standard valve shape does leave something to be desired, however, and the actual valve seat contact area is almost always too large. The way to successfully reduce the valve seat contact area is to regrind the valves with a back chamfer and this is done in a valve refacing machine. Even brand new standard valves need to be reground in this area.

The valve seats in the cylinder head will need to be accurately reground by an engine reconditioner/engine machine shop. The valve seats must be reground until they are totally clean, with no pit marks or distortion of any description on them. Regrinding all valve seats is strongly recommended just to be sure that the seats are concentric with the valve guide and in perfect condition. You cannot judge a valve seat from appearance alone: although a seat may look okay, often it's not!

It may well look like it is possible to lap the valves into the seats and clean the seats up this way, but this is not recommended. Lapping valves into new seats is recommended, but only by a minor amount just to make sure that the valve seats are correct. When correctly ground valves are lapped in, full contact is seen immediately. Lapping is done using fine lapping paste only (400 grit mould makers paste is ideal, although the fine paste readily available at automotive supply outlets is actually coarser it is quite acceptable), never use coarse or medium. If the valve seats need anything more than a minute or so of lapping, then there is something wrong

CYLINDER HEADS

This 1.875in diameter inlet valve has been 'back chamfered' at an angle of 30 degrees in a valve refacing machine. The actual valve seat width is now 0.060in.

with the seat contact area of the valve or the seat contact area in the cylinder head, or both. Avoid prolonged lapping of valve seats.

It is just not possible to successfully lap valves and seats that have something wrong with them (pit marks, deep radial lines, non concentric seat). Even though it may seem that after an interminable time the seat starts to look quite good and acceptable it will never ever be right or

SPEEDPRO SERIES

The 1.75in diameter inlet valve on the left is standard. The inlet valve on the right has been 'back chamfered' in a valve refacing machine.

last anything like as long as a valve and seat that have been correctly ground and then lightly lapped together. When the valve seats are not perfect compression seal is lost and engine power is reduced.

A perfectly smooth finish on the seat contact area between the valve and the seat insert of the cylinder head is the only acceptable condition.

CYLINDER HEAD (ALL TYPES) - REFACING

Any Jaguar XK cylinder head that is being used in a racing application should be refaced, just to make sure that there is no warpage of the head gasket matching surface. The time to have the cylinder head skimmed is after all of the porting work and valve seat re-grinding work has been done. This avoids the possibility of damaging the gasket surface of the cylinder head after it has been skimmed to clean it up. In fact, it's a good idea to skim these cylinder heads whenever the engine is down for a rebuild. Remove as little material as possible in this instance.

To clean up an old, well used cylinder head fully may mean taking as little as 0.005in/0.127mm off. XK cylinder heads can, in fact, be planed by quite a lot to increase the compression. The removal of 0.040/1.106mm from the cylinder head gasket surface to increase CR is quite acceptable and still leaves several re-skims available for the future. This means that over the life of the cylinder head, 0.060in/1.524mm can be removed from the cylinder head gasket surface and the cylinder head will still be sound. There is plenty of material on the gasket surface of these cylinder heads. However, the gasket surface of the XK Jaguar cylinder head is structural, and taking off of larger amounts of material is not recommended. **Caution!** - The cylinder head gasket surface must be absolutely dead flat.

TAPPET GUIDES (ALL HEADS) - SECURING

These are held in the cylinder head by an interference fit. Unfortunately, these

CYLINDER HEADS

This exhaust valve has been 'back chamfered' at an angle of 30 degrees in a valve seat refacing machine. The actual valve seat width is now 0.080in.

SPEEDPRO SERIES

Both of these valves have been lightly lapped.

guides have been known to move at times resulting in a damaged engine. Provided they are soundly fitted (not loose already) they can be prevented from moving by locking them in place using grub screws. This is a good precautionary move. The grub screws will prevent the tappet guides moving up and down in the cylinder head if they become loose. There is one grub screw per tapped guide and each screw hole is drilled from the sparkplug side of the cylinder head.

Opposite - The exhaust valve is often slightly shrouded by the aluminium of the combustion chamber: this material needs to be removed (where arrowed) for the full 360 degrees around the exhaust valve.

www.velocebooks.com/www.veloce.co.uk
All books in print • New books • Special offers • Gift vouchers • Forum

CYLINDER HEADS

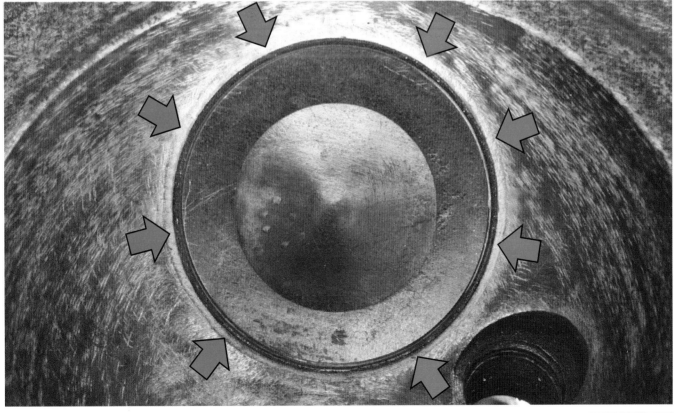

SPEEDPRO SERIES

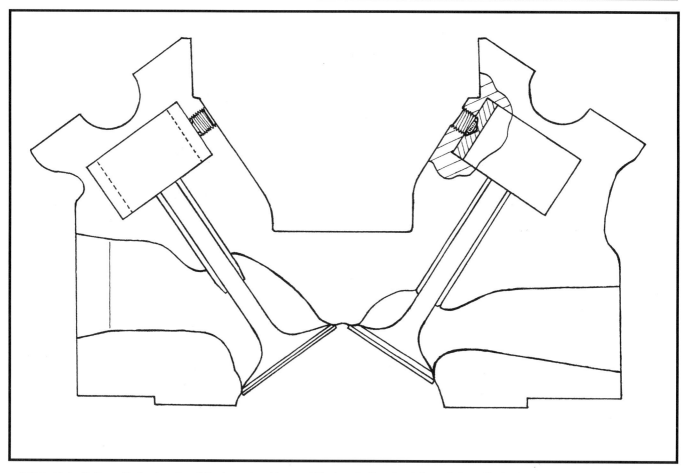

Although both the cylinder head and the tappet guide are drilled, only the aluminium of the cylinder head is tapped. Use a locking agent on the threads and do not overtighten the grub screws as this could distort or crack the tappet guide.

Chapter 4
Camshafts & valve springs

The standard camshafts are quite satisfactory for the standard road-going engine, but they do limit the engine's performance once the cylinder head has been modified to improve gas flow. The XJ6 engine's camshafts are the best of the standard type, being that bit more efficient than earlier standard cams. Although the standard camshaft lobes are very pointed, they do actually offer reasonable lift and reasonably rapid valve action as they sweep across the tappets, though it would be difficult to make the earlier camshafts milder in their action.

Fitting competition type camshafts without cylinder head work does not improve the engine power much, although there is some small improvement accompanied by an increase in fuel consumption. This reflects the fact that improving XK engine performance is mainly about headwork and improving airflow through the ports.

For in-depth information on camshafts and camshaft timing there is another book in the SpeedPro range that deals specifically with this subject - *How To Choose Camshafts & Time Them For Maximum Power* (also by Des Hammill).

Caution! - When very long duration, rapid action camshafts are fitted to XK engines, the prospect of the valves colliding with each other is a very real one. Valves can even get bent when assembling the engine and timing it. On some engines, with long duration camshafts and lots of valve lift, turning the engine backwards with the chains not tensioned properly can be enough to cause the valves contact each other! The only solution is to remove the chains and sprockets and start again. Camshafts within the recommended specifications are not likely to cause problems, but it's possible, so great care must be taken.

CAMSHAFT CHOICE - ROAD GOING

The duration of the later standard camshafts for both inlet and exhaust valves is the same at 256 degrees, and both camshafts give 0.375in/9.52mm lift. The phasing of these camshafts is 17-59-59-17. That means the inlet valve opens 17 degrees before top dead centre (BTDC) and closes 59 degrees after bottom dead centre (ABDC). The exhaust valves are opened 59 degrees before bottom dead centre (BBDC) and closed 17 degrees after top dead centre (ATDC).

The ideal specification for a high performance camshaft will increase the duration over standard, while maintaining the standard valve lift of 0.375in/9.52mm or increasing it to 0.430in/10.92mm. Camshafts of this type will have from 260 to 270 degrees of duration, with 30-60-60-30 phasing. That's the inlet valve opening 30 degrees BTDC and closing

SPEEDPRO SERIES

60 degrees ABDC, and the exhaust valve opening 60 degrees BBDC and closing 30 degrees ATDC. With this type of camshaft the engine will have a hint of idle roughness, but it will be acceptable. The more modified the cylinder head, the more pronounced the idle roughness will be. A standard cylinder head engine equipped with these sorts of camshafts will have a reasonably smooth idle.

CAMSHAFT CHOICE - RACING

The first stage of alternative camshaft types for racing will feature valve lift up to 0.450in/11.43mm (seldom more). Duration will be from 280 through to 290 degrees, or perhaps a bit more on some camshafts. Consider 295 degrees to be the maximum amount of duration.

Expect a 280 degree duration camshaft to have phasing of 35-65-65-35. That's opening the inlet valve 35 degrees BTDC and closing it 65 degrees ABDC and opening the exhaust valves 65 degrees BBDC and closing them 35 degrees ATDC.

Cams with around 290 degrees of duration will have phasing of 43-67-67-43. That's the inlet valve opening 43 degrees BTDC and closing 67 ABDC and the exhaust valve opening 67 degrees BBDC and closing 43 degrees ATDC.

Consider the longest recommended camshaft duration to be 295 degrees, with valve lift at 0.450in/11.43mm. This means having the inlet valve opening 45 degrees BTDC and closing 70 degrees ABDC and having the exhaust valve opening 70 degrees BBDC and closing 45 degrees ATDC. The phasing of the timing events being 45-70-70-45. This sort of camshaft timing is the limit for the average modified XK engine.

A good example of a racing camshaft which definitely works well in any XK engines prepared for racing, is the Piper JAG6 BP300. This camshaft has 282 degrees of duration and 0.450 inches of lift. The phasing is 37-65-67-35. That is, the inlet opens 37 degrees BTDC and closes 65 degrees ABDC while the exhaust opens 67 degrees BBDC and closes 35 degrees ATDC. Don't be fooled into thinking that because the duration is 'only' 282 degrees, that this camshaft is mild: in fact, it's 'a mover.' The idle's rough, but with good power available from 2800rpm through to 6000rpm plus.

Piper RS Ltd make excellent camshafts for XK engines. Piper's catalogue, though, might appear a little bit confusing because the durations listed are basically the same. The reason for this is that the racing camshaft profile is more rapid in action (opening and closing). That essentially means that while all of their camshafts open at the same time and close at the same time, the racing camshaft lifts the valve off its seat to full lift quicker, keeps the valve open at full lift longer, and closes the valve faster than the recommended road going camshaft. Their road going camshaft, by comparison, opens the valve more slowly, doesn't keep the valve at full lift for as long, and shuts the valve more slowly. Both camshafts are correct for their respective applications.

For any XK engine that is not going to be revved to more than 6000rpm (and that's most of them) the camshaft duration does not need to be any more than 280 to 295 degrees. These engines have very long strokes and very long connecting rods: they do not need to be revved excessively to be effective racing engines. With the XK unit it's more case of keeping the rpm within reliable limits and making the engine as efficient as possible over as wide an rpm band as it's possible to do. Over camming one of these engines is a mistake: a mistake that has been made quite often ...

Racing cams for these engines can be defined as having between 280 to 295 degrees of duration, and valve lift of between 0.430 and 0.450 inches. A high performance camshaft has a rapid

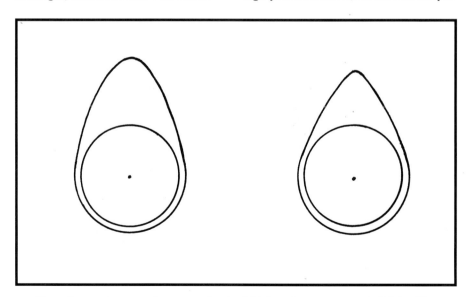

It's quite easy to recognise a non-standard, high performance camshaft lobe. The camshaft lobe on the left is a high lift, 290 degree one, whilst the lobe on the right is standard.

CAMSHAFTS & VALVE SPRINGS

lift action and, as a consequence, has a rounded looking lobe compared to a standard cam's. The difference is quite easy to see with the naked eye.

For racing purposes choose camshafts with durations that suit your particular style of driving. If you use high rpm (gear changes at 6000rpm all of the time) use the 295 degree duration camshafts. If you prefer to use fewer revs, such as 5500rpm gearchange points, in your application, fit camshafts with less duration.

To make real gains in engine power there's no point in fitting camshafts with less than 280 degrees of duration and, within the constraints of the mechanical strength of the standard engine componentry, there's no point in having more than 295 degrees of duration. Within these duration constraints there is still considerable camshaft choice because of the wide range of different camshaft profiles. Provided the camshaft action is rapid the engine will perform well. Look at the actual camshaft lobe itself to judge how good the camshaft will be in the engine, and not necessarily at the advertised duration figures.

As a point of interest, the original D-Type and C-Type camshafts were not particularly high in lift or long in duration. They were, in fact, all within the range of specification suggested by this book.

INLET VALVE CLOSING

With such a long stroke and long connecting rod, the inlet valves should not be closed any later than 65 to 70 degrees after bottom dead centre (ABDC) on the firing stroke. This allows plenty of degrees of crankshaft rotation to build up cylinder pressure before ignition occurs. Given their capacities, these XK six cylinder engines produce excellent torque from virtually just off idle. It is this factor which makes these engines relatively insensitive to increased compression ration (they go well with seemingly low compression).

EXHAUST VALVE OPENING

A common factor with all the aftermarket camshafts meeting the specifications recommended by this book, is that the exhaust valves are not opened too early. The earliest the exhaust valves are opened is 70 degrees BBDC. If the exhaust valves are opened any earlier, engine torque is lost. The one thing these XK engines have in spades, is torque. By opening the exhaust valve any earlier engine efficiency is lost. If these engines were capable of being revved higher than 6000rpm reliably the situation would be slightly different, the exhaust valves could then be opened a little bit earlier (75 or even 80 degrees BBDC). Fitting camshafts which open the exhaust valves too soon to an XK engine that spends most of the time revving between 3000 and 6000rpm is not desirable. While the top end power will not be improved, the low end engine response will be poorer. The engine might well sound a bit louder and idle a bit rougher, however it will not perform as well overall.

VALVE OVERLAP

This refers to the number of degrees of crankshaft rotation where the inlet valve and the exhaust valve of a single cylinder are open together. This overlap condition occurs when, just before top dead centre (TDC), the inlet valve starts to open and the exhaust valve is about to close. A certain amount of interaction occurs during this time which causes engine idle roughness and determines the point in the rev range where the engine 'smoothes out' and 'goes.'
If you have fitted racing camshafts to your engine with the durations similar to those recommended in this book and yet there is no definite point in the rev range when the engine really starts to go (come 'on cam'), there is something wrong with the engine. The way to check this is to increase the rpm of the engine very slowly from idle speed and watch the rev counter to see whether at some point the revs suddenly shoot upwards without any extra movement being applied to the throttle pedal. With camshafts of 280 to 295 degrees of duration this should happen around 2700 to 3200rpm. Expect the engine revs to rise at least a further 1000rpm for no further throttle movement. This is in fact the point where the engine 'comes on the cam.'

HIGH PERFORMANCE CAMSHAFTS - SOURCES

There are five options here –

The first option, and least likely, is to buy a pair of second-hand, but genuine Jaguar factory made D-Type camshafts in good condition. While possible, this scenario is unlikely because of the age of the camshafts and the fact that there were not all that many pairs available in the first place.

The second option is to buy a pair of new aftermarket camshafts from a camshaft specialist which has ground the profiles on to new cast iron blanks or steel billets (very expensive) or standard camshafts which they've built up. Camshafts made like this have standard sized, or near standard sized, base circles and use standard Jaguar shims for valve clearance adjustment.

The third option is to have your own camshafts' lobes built up using the white grade of 'Colmonoy' (Stellite) or a similar product, and then have a new performance profile machined onto the 'new' larger lobes by a camshaft regrinder using a suitable profile. The advantage of this method is being able to use standard Jaguar adjustment shims, and have a camshaft which is not undercut. This is a very common way of creating good high performance camshafts and, although quite expensive, they could literally last forever. Most camshaft regrinding businesses will be able to undertake this sort of work.

The fourth option is to buy a second-hand pair of 'third option' camshafts. The thing about camshafts built-up with Stellite is that they do not wear out easily at all. Even after many years of use they can still be in near perfect condition. When buying second-hand camshafts in this category check them for straightness and equality of lobe size. There are some bargains to be had because you'll find, in many instances, that these cams have been used unsuccessfully (because the engines were not tuned properly and their full potential never realised) in the original application and when the camshafts are removed they're often sold to recoup some money (a bit like the Weber situation). Expect to pay good money for a pair of good second-hand camshafts.

The fifth option involves grinding a new high performance profile onto a pair of existing, late model, camshafts that have not been built up. This will mean that the original camshaft will be seriously undercut, feature a base circle diameter that is reduced from standard, and will not be able to use standard Jaguar shims for valve clearance adjustment. This problem will necessitate having to make up special shims which are much thicker than the standard Jaguar shims.

This last option is the least expensive way of getting a high performance camshaft but it does create two problems. The first is that the camshaft is not as strong as a standard camshaft because it has been undercut (not really good engineering practice). Breakage, although rare, is possible, especially if the undercutting results in sharp corners being formed on the core. The second problem is that thicker shims always have to be made up to set the valve clearances. These shims are made by turning new shims on a lathe (out of high tensile steel), hardening them, and then surface grinding them to the required size. A precision general engineering works or toolmaking company can do this work quite easily, but always get plenty of extra shims made up.

Caution! - XK camshafts that have simply been undercut and then reground (as in 'option five'), never look particularly nice because the base circle has to be undercut by so much. Amazing as it may seem, though, camshafts treated like this seldom ever break in service. No camshaft grinder is ever going to guarantee that a camshaft ground like this is not going to break though, so you need to be aware that failure is a possibility, if unlikely.

Some other considerations

Standard XK camshafts can be specially built-up and then reground very successfully to take all manner of performance regrinds by any competent camshaft reprofiling company or engine reconditioning works. Provided the camshaft profile is designed for an overhead camshaft engine which operates through inverted bucket type lifters, it will be suitable. The usual problem is that, while there are plenty of camshaft profiles available for this design of valve gear, most camshaft lobes do not have the required lift (0.400-0.450in/10.16-11.43mm).

The durations recommended are approximations. This means that if the camshaft grinding company has a profile listing that gives timing events which are 37-67-67-37, consider this to be very similar to a pair of camshafts that will give 35-65-65-35 timing. The range of lifts is vital, however, and must be in the region of 0.400 to 0.450 inches. As a general rule, the more duration there is, the greater the lift as a consequence of design considerations.

A bit of 'mixing and matching' is possible when it comes to camshaft choice. It is quite possible to fit an inlet camshaft with 295 degrees of duration and 0.450 inches of lift, in conjunction with an exhaust camshaft which has 285 degrees of duration and 0.400 inches of lift. This will mean that the timing might be 45-70-65-40. That's inlet valves opening 45 degrees BTDC and closing 70 degrees ABDC, while the exhaust valve opens 65 degrees BBDC and closes 40 degrees ATDC. The longer inlet timing will often produce better top end performance, while the reasonable, but different, exhaust timing will ensure that the engine has the maximum possible torque.

Once the exhaust valve opens, of course, the expanding charge is no longer working effectively on the top of the piston and pushing it down. Obviously, if the exhaust valve is opened too late there will not be enough time to clear the cylinder of the burnt gases. No road going engine, therefore, needs to have the exhaust valve opened any earlier than 65 to 67 degrees BBDC, and no racing engine revving to 6000rpm needs to have the exhaust valve opening any earlier than 70 to 72 degrees BBDC.

Finally, XK camshafts are, strictly speaking, right and left handed. Once an exhaust cam always an exhaust cam and once an inlet cam always an inlet cam.

CAMSHAFTS & VALVE SPRINGS

It's a good idea to replace these items with new genuine replacement parts.

VALVE COLLETS AND KEEPERS

Most of these engines will have done thousands of miles by now and most will still have the original collets and keepers fitted. **Caution!** - It is strongly recommended that brand new, genuine Jaguar collets and keepers be bought and fitted to any engine that is being rebuilt for high performance. This is a measure to preclude, as much as possible, the failure of one of these items. Failure of these parts is rare but it does happen from time to time. For the modest cost of brand new replacement parts the benefits can be enormous.

VALVE SPRINGS

Caution! - Always use new valve springs in high performance applications so that the valve spring tension is the maximum possible with that particular spring set. Valve springs lose tension with age and use, so never fit used valve springs to a high performance engine unless they're relatively new and still make their rated poundage (within 5%).

Caution! - Electronic rev limiters are readily available and should be fitted to all XK engines when they are being used in competition. With standard, but new, valve springs fitted there is not a lot of room for error with regard to engine over speeding. The standard poundage, while being more than adequate for a standard engine does not allow all that much of a safety margin at engine speeds beyond 6000rpm. Fitting a rev limiter is the solution to possible valve damage via engine over speeding (caused by missed gear shifts, and so on), where the engine rpm can rocket. Avoid this situation at all costs.

Standard type

The standard dual valve springs are designed to work with standard valve lift and not significantly more. Taken to the safe limit, this means that standard dual valve springs are only good for a maximum of 0.410in/10.41mm of valve lift. That's 0.035in/0.889mm more than standard. **Caution!** - Even though the standard dual valve springs will take slightly more than this, it's not safe on the basis of valve spring reliability (too near to coil bind). Jaguar valve springs are, of course, of excellent quality, but there are limits to just how much a valve spring can be compressed before spring breakage, through fatigue, can happen.

The fitted height of the outer valve spring in XK cylinder heads is approximately 1.325in/33.65mm. The fitted height of the inner valve spring is approximately 1.250in/31.75mm. The coil bind height of later valve springs (in combination) is approximately 0.875in/22.22mm. This means that the valve springs in combination are coil bound at 0.450in/11.43mm of valve lift. This is why the maximum recommended valve lift using standard valve springs is 0.410in/10.41mm as it leaves a total clearance between the coils of 0.040in/1mm.

The seated tension of later XK valves springs, for example, is approximately 70-80 pounds/31.75-36.28kg. The (valve) fully open pressure is approximately 130-140 pounds/58.96-63.50kg at 0.375in/9.525mm of valve lift (outer valve spring compressed to approximately 0.950in/24.13mm). These later standard dual valve springs are quite sufficient for all high performance engines with camshafts featuring up to 0.410in/10.41mm of valve lift and engines that are rev limited to 6000rpm.

Caution! - Irrespective of what valve springs you buy, they should all be checked for installed height pressure (at 1.325in/33.65mm), and the coil

SPEEDPRO SERIES

Standard dual valve spring.

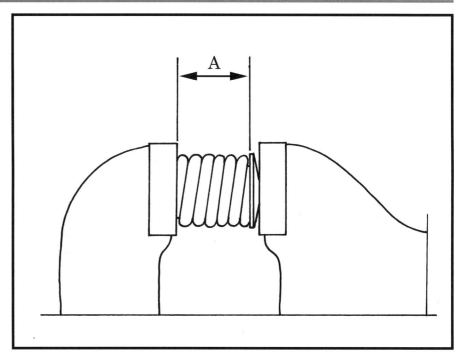

Both coil springs are fitted and the valve retainer is in place. The springs have been compressed until they are lightly coil bound. A vernier calliper is then used to measure the compressed height of the coil springs as a pair. One of the springs will become coil bound before the other (usually the inner one).

bind height of the two valve springs should be ascertained by squeezing them in a vice with the valve spring retainer fitted on top of the springs (Warning! - wear eye and body protection). Whatever camshaft lift you are going to use, the valve springs must (from installed height) compress by the amount of lift plus a minimum of a further 0.040inches/1mm before they are coil bound. To reiterate, that's the fitted height of 1.325in/33.65mm minus the valve lift, and minus a further 0.040in/1mm. Taking the new valve springs out of their packaging and just fitting them, and hoping for the best is pure folly ...

In summary - If standard valve springs are to be used, the maximum camshaft lift recommended is 0.410 inches. Be aware of the fact that many camshaft lobes often do not actually make the advertised amount of lift. Some camshafts are short of the listed figure by between 0.005 and 0.015in/0.12 and 0.38mm. So, in some instances, a camshaft listed as having 0.425 inches of lift might, in fact, only have 0.410 inches or so of lift.

Stronger valve springs

Stronger than standard valve springs suitable for the XK engine are made by a large number of valve spring manufacturers. Catalogues are available from most aftermarket camshaft manufacturers, and they usually list the range of valve springs that they sell. Because of the relative rarity of the Jaguar XK engine these days, most aftermarket manufacturers do not list specific valve springs for use in these engines so, the physical dimensions of the springs and the pressures have to be matched up to what they do sell for other engines. This is not a difficult task provided full details of the valve springs are listed in the catalogue. Many camshaft manufacturers have technical help lines, of course, and contacting a company in this manner will usually lead to you being told exactly what they have available to suit these engines.

Unless the standard 0.375in/ 9.52mm of valve lift is going to be retained, or up to 0.410in/10.41mm, the standard dual valve springs will have to be replaced with alternative dual valve springs. Aftermarket valve springs suitable

CAMSHAFTS & VALVE SPRINGS

for this particular application must be able to be compressed down to 0.835in/21.20mm or, preferably, more. The maximum valve lift likely to be encountered on any suitable aftermarket camshaft is 0.450in/11.43mm, for which the valve springs used will have to be able to be compressed a further 0.040in/1mm more than the true valve lift before the coil springs are coil bound.

Caution! - The coil bind height of the two valve springs should be ascertained by squeezing them in a vice with the valve spring retainer fitted on top of the springs (Warning! - wear eye and body protection). Whatever camshaft lift you are going to use, the valve springs must (from installed height) compress by the amount of lift plus a minimum of a further 0.040inches/1mm before they are coil bound. To reiterate, that's the fitted height of 1.325 inches minus the valve lift, and minus a further 0.040 inches. Taking the new valve springs out of their packaging and just fit them and hoping for the best is pure folly.

Aftermarket high performance dual valve springs will have more tension, but they don't need to be excessively strong. Consider 180 pounds/81.64kg of (valve) fully open valve spring pressure to be the absolute maximum to use (revs limited to 6000rpm). The new valve springs must have at least 160 pounds/72.57kg of full lift ('over the nose') valve spring pressure, for up to 6000rpm to be used with aggressive action camshafts.

Caution! - Valve springs that are taken to near coil bind each time the valve is opened have a much shorter fatigue life and, further to this, valve springs often get firmer in action when they get within 0.020in/0.50mm of coil bind and this can lead to excessive camshaft lobe 'nose' wear.

VALVES SPRINGS - CHECKING

Caution! - Check all valve springs for poundage (pressure exerted at a given height) and replace any valve spring combination that fails to measure up to its maker's specifications. Checking valve springs for pressure is not all that difficult. It can be done using a set of bathroom scales and a drill press. Bathroom scales are not necessarily all that accurate, but they generally give a consistent reading.

New aftermarket valve springs are usually supplied with relevant specifications (which should be noted for future reference) while original equipment springs should match Jaguar's specifications.

The importance of valve spring pressure should never be underrated. If a valve is controlled by springs which do not have enough tension, that valve can 'float' (the spring doesn't react quickly enough to close the valve properly) at high rpm, which could lead to the destruction of the valve (the head falling off, for example). Even a brand new valve can only stand so much of this sort of treatment.

Most engine reconditioning workshops or machine shops will be able to measure spring tension for you. Alternatively, the bathroom scales and drill press method while not so accurate will certainly pick up a valve spring combination which has a different tension from all of the other valve springs in the set. This procedure is fully covered in the SpeedPro book *How To Choose Camshafts & Time Them For Maximum Power*.

Any increase in valve lift requires valve springs that allow more valve spring compression without coil binding. The fitted height of the average XK engine's valve springs is 1.325in/33.65mm. Find and fit dual valve springs that do not exert more than 160 to 180 pounds/72.57 to 81.64kg of pressure at the full lift height of the camshaft lobe (measured from the spring's fitted height).

With, say, a 0.450in/11.43mm valve lift, that means checking to make sure that the valve spring pressure is not more than 180 pounds/81.64kg at a compressed height of 0.875in/22.22mm. If the valve springs register 150 to 160 pounds/68.04 to 72.57kg of pressure, for example, that will still be sufficient for 6000rpm operation. In most instances, the dual valve springs that are available will show more than 150 to 160 pounds/68.04 to 72.57kg of pressure at a compressed height of 0.875in/22.22mm.

Once the actual camshaft lift to be used is known, the amount of lift can be deducted from the fitted height of 1.325in/33.65mm and the valve spring combination checked for coil bind height by placing each pair of dual valve springs in a vice and taking them to coil bind by closing the vice up and measuring the closed height with a vernier calliper (Warning! - wear eye and body protection). **Caution!** - The valve springs must be able to be compressed at least 0.040in/1mm more than the full camshaft lift height.

The outside diameter of a standard XK outer valve spring is 1.125in/28.57mm. The inner valve spring has an inside diameter of 0.630in/16mm. Any valve springs used must be as near to the standard Jaguar valve spring diameters as possible since alternative valve springs must fit the standard valve retainer properly.

The free standing height (free length) figures of the standard valve springs vary a bit and are not particularly relevant. What is relevant, however, are the coil bind heights of both valve springs, the fitted height valve spring pressures and the full lift valve spring pressures.

SPEEDPRO SERIES

Warning! - Observe all safety precautions to avoid personal injury and injury to others when valve springs are compressed. They store a lot of energy and, if accidentally released, can cause serious injury/damage on impact.

Caution! - The valve retainer must be in place on the valve springs when they are compressed to measure coil bind height or pressure. This is because the inner valve spring is retained at a different height than the outer valve spring and is, as a consequence, compressed more. A false reading, which could lead to engine damage, is possible if the retainer is not in place. The standard XK valve spring platforms on which the valve springs sit in the cylinder head are flat.

The retainer and the two collets on the left are later, 'short' ones. Those on the right are early, 'long' ones.

VALVE SPRING RETAINER/ OIL SEAL - CHECKING CLEARANCE

If the valve lift of your chosen camshaft is more than 0.425in/10.79mm, a further consideration has to be taken into account. Later XK cylinder heads have factory fitted valve stem seals on the inlet valve guides, as opposed to the 'taper topped' valve guides (no seal) as found on earlier heads. **Caution!** - These seals will be contacted by the underside of the valve spring retainer if the valve lift is more than 0.425in/10.79mm. The solution to this problem is to remove the seal (which will then allow approximately 0.500in/12.7mm of valve lift) and ensure that the valve guide to valve stem clearance is as per the minimum specified factory size.

Any earlier 'taper topped' valve guide equipped cylinder head has approximately 0.530in/13.64mm of clearance between the valve retainer and the top of the valve guide when the valve is seated. Earlier engines (B-type for example) also have deeper valve spring retainers and different collets to the later XJ6 engines.

Caution! - Make sure that you have the right valve retainers and collets for the particular cylinder head.

With the valve stem seals simply left off the tops of the inlet valve guides, there is the prospect of the engine burning oil, or to be more specific, oil being drawn down the valve guides and contaminating the incoming air/fuel mixture. If the valve guides are restored to as-new condition with K-Line valve guide inserts with the factory recommended minimum clearance, there will be virtually no oil contamination. What does happen, though, is that the engine will often give one puff of oil smoke when it is started, but this is acceptable. If there are large valve stem to valve guide clearances, too much oil will be able to go down the guides and the engine will be an 'oil burner' and lose power because of it.

The exhaust valve stems do not have valve stems seals fitted to them as standard and, as a consequence, there is no full lift clearance problem. There is approximately 0.600in/15.24mm of clearance between the valve retainer and the top of the exhaust valve guide at zero valve lift.

To avoid any interference problems between the underside of the valve spring retainer and the top of the inlet valve guide seal, check each combination. This is done by fitting the inlet valves to the cylinder head and the spring retainers with their collets (no valve springs fitted). With the valve held manually in the closed position (from the combustion chamber side of the cylinder head), measure the distance from across the top of the tappet sleeve down to the top of the valve spring retainer using the tail of a vernier calliper. Then push the valve to the fully open position and measure the distance again. The difference between the two measurements is the maximum distance the valve can move before the underside of the retainer contacts the top of the seal. Check each valve in this manner just to make

CAMSHAFTS & VALVE SPRINGS

Full lift timing points for inlet and exhaust camshafts are permanently marked on this crankshaft damper.

sure that sufficient clearance exists.

Fitting valve stem seals to the inlet valves is recommended but, sometimes, because of high lift, it is not easy to achieve.

CAMSHAFTS - TIMING

Camshaft timing is all important if an XK engine is to go really well. Although an engine can have all the 'best bits' money can buy, it won't perform as well as it might if the camshaft timing is not set up as per the camshaft manufacturer's specifications. This sort of attention to detail (checking and then setting the camshaft timing) can make the difference between the engine being an also ran or being a winner.

In the first instance, make sure that all of the camshaft timing details are with the camshafts when you buy them. When camshafts are bought new, the manufacturer or camshaft grinder will include, as a matter of course, all the relevant timing details. Secondhand camshafts should still come with written details of some sort and, failing that, most camshaft regrinders stamp their code on camshafts to identify them. Consequently, provided the manufacturer or regrinder is known, the details can be requested.

As an absolute minimum you simply must have the tappet clearances and the full lift timing positions of each camshaft. Going by the standard slot positions in each camshaft can sometimes be enough to get aftermarket camshafts timed 'about right' (within 2 or 3 degrees) but is not good enough for someone who wants optimum performance. There can be manuafacturing error here, so each camshaft must be checked to see if the slots are accurately positioned in relation to the lobes.

Each camshaft must be checked in turn using the 'full lift' position against an accurately degree-marked crankshaft damper, or a correctly positioned degree (timing) wheel. By far the best method is to accurately mark the crankshaft damper permanently with the full lift inlet and full lift exhaust timing points. Usually these points will be around the 105 to 110 degrees after top dead centre (ATDC) for the inlet timing full lift position, and 105 to 110 degrees before top dead centre (BTDC) for the exhaust full lift timing point, but you need the precise figures for your camshafts. This way, the engine always carries appropriate datum marks and there is never a subsequent need to set up a timing disc for checking purposes. The crankshaft damper of an XK engine is quite large in diameter and ideal for carrying these extra markings.

Before anything else, top dead centre (TDC) has to be checked for accuracy (if it isn't dead accurate the engine will never be timed correctly). All camshaft timing and ignition timing relies on the fact that the top dead centre position is proven to be exactly right (see chapter 5 for checking top dead centre detail). Inlet full lift and exhaust full lift degree points are clearly and permanently marked on the damper in relation to the proven TDC mark.

Use the standard slots in the camshafts and the standard key plate to initially time the engine and then

SPEEDPRO SERIES

Degree point has been filed into the rim of the damper and the numbers and letters have been stamped and then paint filled to highlight them.

check the full lift timing against the degree markings.

If the camshaft timing is really well out, an engine can be down 30bhp or more and, worse still, not have anything like the 'snap' (acceleration) that it would if the camshafts were timed correctly.

A dial test indicator (dial gauge) with a long spindle, long travel and a magnetic stand will be necessary to do this work. Dial test indicators can be expensive, but there really is no other way of detemining the full lift position quickly, easily and really accurately. Because the cylinder head is aluminium a steel plate will have to be bolted onto the cylinder head and the magnetic stand fixed to that. An alternative here is to make a special fixture that clamps the dial test indicator in position, but it is generally easier to temporarily bolt a steel plate onto the cylinder head. It is easier to position the dial test indicator using the freedom of movement that the magnetic standard offers.

Using the full lift position means picking up that point on the 'toe' of the camshaft lobe between where the valve just reaches full lift, dwells (the middle of the dwell is the true full lift position) and then starts to descend. This means that even if the tappet clearance is set incorrectly, or the camshaft opening and closing points are not as accurately ground as they should be, or the camshafts have been reground and the operator of the camshaft grinder was not able to align the original factory machined slots, these factors will have no effect on the full lift timing point position. The full lift timing position is a clear cut point, with little room for error. This is why using full lift is an ideal method of checking and setting the camshaft timing on one of these engines.

A further point here is that most high performance camshafts for XK engines have relatively late exhaust valve opening and early inlet valve closing. This means that the exhaust valve opening point, for example, is never going to be more than 70 degrees before bottom dead centre (BBDC) anyway, and the possibility of a major loss of engine torque through exhaust opening too early just doesn't exist. There is usually at least 110 to 115 degrees of crankshaft rotation before the exhaust valve opens. The same goes for the inlet closing point which is usually never later than 70 degrees after bottom dead centre (ABDC). There is usually 110 to 115 degrees of crankshaft rotation on the compression stroke. Many other designs of engine are more sensitive to the exact exhaust opening degree point and, to a lesser extent, the exact inlet closing degree point, but the XK engine is not. There are other methods of timing camshafts but none seem to be better for the XK unit than the 'full lift' timing point method recommended here.

If, after checking the full lift timing positions of each camshaft, the camshaft timing slots are found to be exactly right, then the factory gauge can be used with confidence to set the camshaft timing from here on (but only if the slots are right).

When setting the camshaft timing on one of these engines, the timing chains must be tensioned correctly. Once the tension is set, the adjuster needs to be marked so that if the chains need to be slackened off during the camshaft timing procedure, the tensioner can be repositioned in exactly the same place. This is because the amount of tension and, more specifically, the position of the tensioner, can affect the timing of the inlet camshaft (less tension will retard the inlet camshaft's timing).

Caution! - It must be possible to turn the engine over freely (must have piston to valve clearance) and it is strongly recommended that the engine be turned clockwise (looking at the crankshaft damper from the front of the engine) only. If it has been built

CAMSHAFTS & VALVE SPRINGS

Exhaust camshaft being timed in the full lift position.

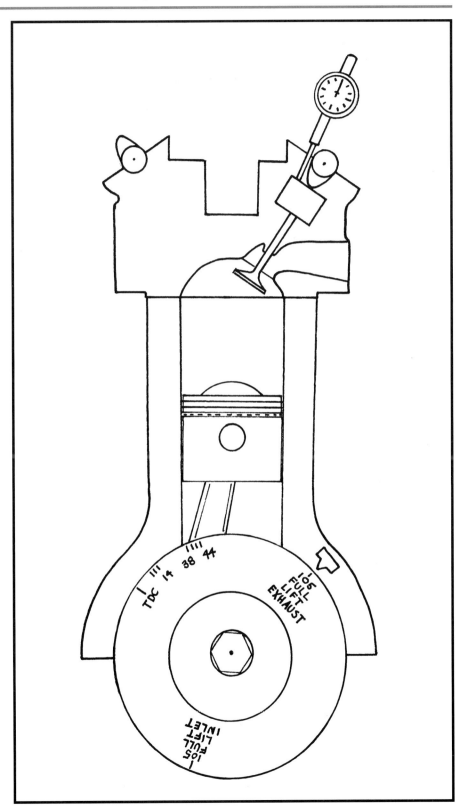

SPEEDPRO SERIES

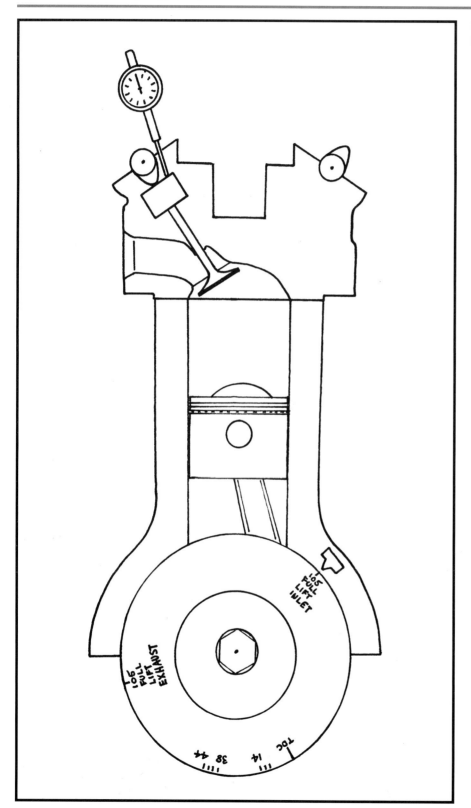

Inlet camshaft being timed in the full lift position.

CAMSHAFTS & VALVE SPRINGS

correctly, turning one of these engines over with the sparkplugs out is not all that difficult. If there is any resistance to turning the crankshaft, check for piston to valve interference. It is very easy to bend a valve or two at this point through incorrect valve timing. The 'wilder' the camshafts, the more careful you have to be. If the engine appears to be 'locked' (meaning you can't turn the crank forwards or backwards), take off the camshafts and start again. DO NOT be tempted to turn the engine past the tight spot as bent valves will be the result if piston to valve interference is the cause of the resistance. Setting the camshaft timing on these engines is a tricky business and patience is definitely required.

The exhaust camshaft is always timed first. This is because the exhaust camshaft is 'before' the chain tensioner. Once the exhaust camshaft has been checked for timing and, if necessary, adjusted it does not need to be looked at again. With the dial test indicator positioned on the number one cylinder's (the rearmost cylinder) exhaust valve follower, the crankshaft is turned so that the exhaust valve is open and lifting. You can, of course, use the number six cylinder (the front cylinder) to check camshaft timing and it is often more convenient to do so.

As the valve approaches full lift (clearly seen by the fact that the 'toe' of the camshaft is in the middle of the follower) the dial indicator needle will be seen to slow in its movement and then stop. The moment this happens, stop turning the crankshaft and see where the full lift degree position (as marked on the crankshaft damper) is in relation to the timing pointer and temporarily mark the damper/pointer relationship with a pen. The crankshaft is then turned further (about 2 to 3 degrees) until the moment when the dial test indicator needle just starts to move again (valve starts to close). Stop turning the crankshaft the instant the dial test indicator needle starts to move. Look to see where the full lift timing line, as marked on the damper, is in relation to the timing pointer and temporaily mark the timing pointer position with a pen.

The full lift timing point should be exactly between the two temporary pen marks you made. The distance between the two temporary marks represents the timing dwell which will vary from cam design to cam design, but will usually represent only two degrees of crankshaft rotation.

While the recommendation is to use the camshaft manufacturer's full lift timing points, enthusiasts need to be aware that there is more to camshaft timing than this if optimum engine performance is to be realised. While the accuracy of camshaft lobe phasing is more or less guaranteed, engines very often respond to having the camshafts set in slightly different positions, such as 1-4 degrees advanced or retarded of the factory prescribed positions. It takes some work to time the camshafts of Jaguar engines, and there is the ever-present problem of bent valves if you get it slightly wrong. Jaguar camshafts tend to get timed to the manufacturer's recommended positions, and then left as they are. This is almost always very acceptable as, in most cases, setting the camshafts to the manufacturer's settings will be good enough to achieve excellent power. However, you might make some further gains in efficiency by setting the camshafts in different positions and testing the acceleration rate of the vehicle.

Changing the camshaft timing a couple of degrees can sometimes result in a performance improvement (acceleration rate). In other words, you find the 'best' position for the camshafts you have in your engine (in terms of optimum maximum power and fastest acceleration rate of the vehicle) by testing. It's a pity that it takes so long to make a change to the camshaft timing on the XK engine. On belt drive engines fitted with Vernier timing wheels, for example, it takes no time at all to check all possible combinations and settle on the optimum settings. A combination of rolling road maximum power tests with different camshaft positions, followed up by road/track acceleration testing over a measured distances soon sorts out what the optimum settings are. You only need to do this sort of testing once to finalise settings. All future engine rebuilds require the camshaft timing to be set to the new full lift timing degree position points, or to the factory recommended full lift timing figures if they did indeed prove to be right.

Chapter 5
Ignition system

There have been two basic models of Lucas distributor fitted to XK engines. The early one had a larger cap and a more rounded body than the later model; it is quite easy to recognise. The later design has been around since the early-1960s, evolving over the years to the point of being electronic as opposed to having contact breaker points.

Contact breaker points type distributors are quite adequate for any of these engines (up to 6000rpm). The later electronic ignition systems, or a points to electronic conversion, are more reliable, though, because there are no contact breaker points to deteriorate or to require changing. Only ever use the standard recommended coils or Lucas Sports Coils, as these are more gentle on point contacts.

The reliability of the electronic distributor makes it ideal but, if your distributor is a contact breaker points type, retaining it will not necessarily result in a reduction in engine performance (assuming the distributor is in good condition, of course).

There is another book in the SpeedPro Series that deals specifically with distributor type ignition systems and will provide further information for those that require it. This book is called *How To Build & Power Tune Distributor Type Ignition Systems* (also by Des Hammill).

For a distributor ignition system to work at its best, the distributor cap, rotor, points, condenser, coil (a Lucas Sports coil or an equivalent is ideal) and the HT leads (plug wires) need to be brand new. HT leads should have a low resistance (5-10kV rating, maximum) or, for competition purposes, be copper cored (virtually no resistance).

CONTACT BREAKER POINTS
If applicable, keep a spare set of new contact breaker points for immediate replacement since these are the weakest link in the ignition system. Set the points gap to 0.014in/0.35mm and do not use a feeler gauge to set points or file the points to clean the actual contacts. The feeler gauge blade will often contaminate the point contacts as it passes across the surfaces. This will cause the engine to misfire. Filing points is not recommended at all and can result in an engine that misfires chronically.

To set the contact breaker point gap, place the feeler gauge next to the point contacts and use the size of the feeler gauge blade as a guide rather than a direct measurement tool. Use the smallest gap recommended by Lucas (0.014in/0.35mm). If the contact surfaces of the points are badly pitted or worn the points set should be replaced, as should the condenser. If the points are simply worn, but not badly pitted, or one set of contacts is pitted more than the other side, its enough to replace the points set alone, in fact it's desirable to retain the

IGNITION SYSTEM

condenser and not damage it as the 'new' condenser might not be as good.

The use of an electronic distributor is highly recommended for any racing engine.

VACUUM ADVANCE

For road use the vacuum advance needs to be retained to ensure that the best fuel economy is achieved. Essentially, the vacuum advance system advances the ignition timing when the inlet manifold is under vacuum (engine has only partial loading) by as much as 10 to 15 degrees over the centrifugal advance. When the engine is under heavier load or is accelerated, the vacuum advance mechanism ceases to operate. For the acceleration aspect of an engine's performance, the vacuum advance system can be disregarded.

For a pure racing engine, the vacuum advance mechanism should be removed from the distributor and the baseplate locked with braze to prevent any movement (which could cause changes in ignition timing). The hole in the body of the distributor, through which the vacuum advance once fitted, must be blocked off to prevent dirt and grime getting into the distributor. It can be be filled by an aluminium plug or with a plastic metal compound (JB Weld, for instance).

STATIC ADVANCE

Static advance is set by positioning the distributor in accordance with timing marks on the crankshaft damper so that the spark plug in number one cylinder fires an appropriate number of degrees before the piston reaches TDC. Crankshaft damper rims are factory marked, in 2 degree increments, from top dead centre (TDC) to 10 or 12 degrees before top dead centre (BTDC).

Standard XK engines, depending on model, use between 4 to 12 degrees of static advance. It is usual for any XK engine to require 12 degrees of static advance. Some engines equipped with long duration camshafts might require 14 degrees.

Static advance is used to roughly time the engine so that it can be started. Once started, the idle speed advance can be set accurately using a strobe light.

Whether static and idle speed advance are the same or different depends on the advance mechanism spring tension.

CENTRIFUGAL ADVANCE

The centrifugal advance mechanism built into the distributor is what causes the ignition timing to advance further as revs rise from idle speed. The total amount by which the distributor can advance the ignition timing is built into this mechanism, and the rate of that advance is controlled by the two advance weight springs (under the points' baseplate).

After a certain point in the rev range has been reached, the total amount of centrifugal advance built into the distributor will be 'all in,' and no further advancing of the spark via the centrifugal advance mechanism can take place.

If the idle speed advance amounts, the centrifugal advance rate, and the total amount of advance are not right, no XK engine will accelerate as well as it could. This aspect of ignition timing is often disregarded, and power is 'lost' because of it. The whole ignition advance system is a vital aspect of XK engine tuning.

PRE-IGNITION

If one of these engines is going to pre-ignite, it is going to do so at maximum torque under full loading. This is because maximum torque is produced when the cylinders have maximum charge density (the maximum amount of air/fuel mixture has been inducted into the engine). On these engines the ignition timing is set to match this situation. As the revs rise above maximum torque the charge density reduces (volumetric efficiency falls off) and the predisposition to pre-ignition reduces. Pre-ignition will, of course, destroy an engine! To a certain degree the ignition advance becomes less critical after maximum torque has been developed.

DISTRIBUTOR MODIFICATIONS

Standard distributors need to be modified for high performance applications because both the rate and total amount of advance will be wrong. Fortunately, the appropriate modifications can be achieved quite easily.

The two standard centrifugal advance springs are usually not suitable for a performance orientated engine. At idle, and off idle, it is only what is termed the 'weak' spring that is advancing the ignition. The 'strong' spring only comes under tension when the 'weak' spring allows sufficient movement of the distributor's advance mechanism so that the posts to which it is fitted move away from each other and cause the stronger spring to come under tension too. At this point the advance rate will slow down drastically (the 'strong' spring is usually very strong). These two springs give the distributor a two phase advance curve which, initially, sees the timing advance quite quickly then slow down as the revs increase to 5000rpm or so. This is not the sort of advance curve that a high performance engine requires.

To achieve a more suitable advance rate, the weaker of the two advance mechanism springs is usually retained and the stronger spring changed. Having said that, some 'weak' advance springs are actually too weak, since they are designed for engines which have a very small amount of static advance (the distributor mechanically advancing to a low total quite quickly). These, too, will have to be changed.

As a general rule, the two advance mechanism springs have to have sufficient tension between them to allow the advance mechanism baseplate to start moving and, consequently, advance the ignition timing from around 1400rpm and be fully advanced at approximately 3000 to 3300rpm. Once the crankshaft damper has been degree marked, timing events are relatively easy thing to check. All Jaguars have rev counters and, by slowly increasing engine revs with the strobe light on the crankshaft damper, the point where the ignition starts to advance can be seen; as can the point where the ignition stops being centrifugally advanced. If the distributor is fully advanced at 2500rpm, the springs, collectively, are too weak. Change one of them for a stronger one and increase the spring tension collectively until the total amount of ignition advance is 'all in' at 3000 to 3300rpm.

Ideally, the sooner after 3000rpm the ignition advance is 'all in' the better. **Caution!** - However, you must allow the 'all in' point to occur at higher revs (by fitting progessively stronger springs) if pinking (pinging) occurs under acceleration. An extra 200rpm can make a difference. Consider 3500rpm to be the absolute limit and, if the engine is still pinking, check the air/fuel mixture and then consider reducing the compression.

Caution! - Always use the highest octane fuel you can obtain on a regular basis. Retarding the ignition to prevent pinking (because the compression is too high for the fuel available) is not the way to go because full throttle cannot safely be used when accelerating the engine.

A good source of mechanical advance distributor springs is scrapped (junked) distributors, and not necessarily Lucas units for XK engines. Scrapyards (junkyards), and even automotive electrical specialists, often have a bin of old distributors amongst their aluminium scrap and they'll usually let you take the springs out for free, or a small fee. You could end up with 20 or 30 different 'weak' distributor advance mechanism springs and one of them is bound to be suitable for your application.

Over the years, Lucas has built a range of degrees of centrifugal advance into its distributors for Jaguar, with 13 degrees being the most common for the XK engine. A distributor with 13 degrees of built in advance is the best starting point for a high performance XK engine, though, ultimately, this limit might not prove to be the ideal for your application. You can ascertain the degrees of advance of a distributor by removing the baseplate and checking the degrees of advance number stamped on the advance weights. They range from 9 to 25 (if the whole range of Lucas distributors as made for XK engines is taken into consideration).

An important factor to take into consideration is the fact that the crankshaft turns twice for every single revolution of the distributor. This means that if the distributor has 13 degrees stamped on it this 13 degrees is actually 26 degrees of crankshaft rotation. That's 26 degrees of ignition advance in relation to the crankshaft position - which is where the ignition (spark) advance is measured. So, for example, with 26 degrees of spark advance from the centrifugal advance mechanism, plus static advance of, say, 12 degrees, altogether there is a total possible 'all in' ignition advance of 38 degrees BTDC. If the static advance is 14 degrees, and the centrifugal advance 26, the total amount of advance will be 40 degrees, and so on.

Engine manufacturers do not specify total advance settings, they only give static or idle settings. The rate of advance and the total amount of advance is, of course, built into the distributor. This is all very well for a standard engine, but it is not an ideal situation for a performance orientated engine. There is just too much information left out of the equation and this is what can lead to a damaged engine (over advanced), or an engine that does not produce the amount of power it is capable of producing (a retarded engine). Once the crankshaft damper has to be accurately marked with total advance degree makings, you can measure the total ignition advance, and other ignition events, virtually anywhere (at the race track, for example), provided you have a strobe light.

For road cars, an additional aspect of the ignition advance scenario concerns the built in vacuum advance. If the vacuum advance is taken into consideration for part throttle operation (high manifold vacuum), the total ignition advance could be as much as 53 degrees (in this example total static and centrifugal advance is 38 degrees 'all in' and the vacuum advance is the maximum possible at 15 degrees). This amount of ignition advance is acceptable for cruise conditions and will give best economy.

CRANKSHAFT DAMPER - CHECKING TDC & ADDING OTHER TIMING MARKS
Finding top dead centre
Without an accurate top dead centre

IGNITION SYSTEM

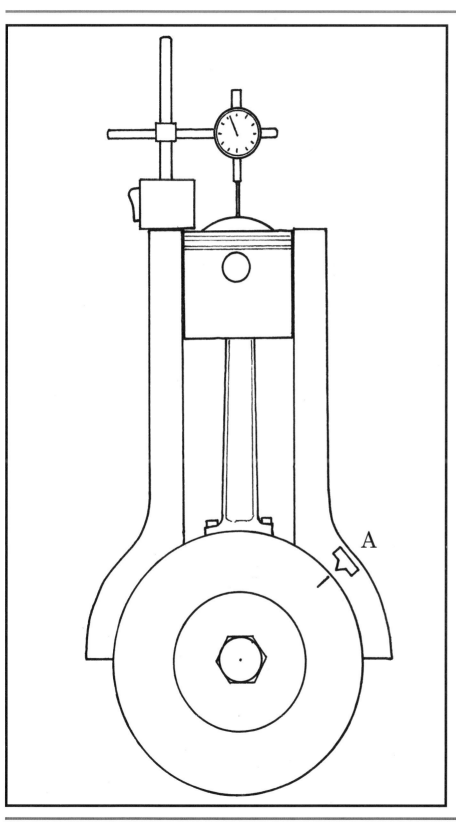

The piston crown is as high as it can go. The damper's top dead centre mark (TDC) is in alignment with the pointer on the block (A).

(TDC) marking no engine can be tuned properly. Jaguar engines are very well made and, invariably, the TDC marks of original engines are accurate. However, because of their age now, and the fact that parts could have been substituted at one stage or another, all engines need to be checked to ensure that TDC is accurately marked on the crankshaft damper.

Degree markings on the crankshaft damper are used for both camshaft and ignition timing. Therefore, when TDC has been determined and the damper removed for the permanent marking of ignition timing degrees on its rim, the camshaft inlet full lift and exhaust full lift timing marks should also be marked at the same time.

Finding true top dead centre requires checking the height of the relevant piston in its bore, usually via the spark plug hole if the engine is assembled. TDC relates to a single cylinder and for XK engines this will be number one cylinder, the rearmost cylinder in the block. The front cylinder can, of course, be used since the front and rear connecting rod crankshaft journals are on the same axis.

The two usual methods of determining TDC involve a 'dead stop' device for contacting the piston crown, or a dial test indicator (dial gauge).

If the cylinder head is off the engine, the dial test indicator method is the easiest to use and the most accurate. A dial test indicator with a magnetic stand is placed on the block and the spindle of the dial indicator positioned to contact the crown of the relevant cylinder's piston. The engine

SPEEDPRO SERIES

An old sparkplug made into a 'dead stop.' The probe of the dead stop contacts the piston crown and prevents it from reaching top dead centre (TDC).

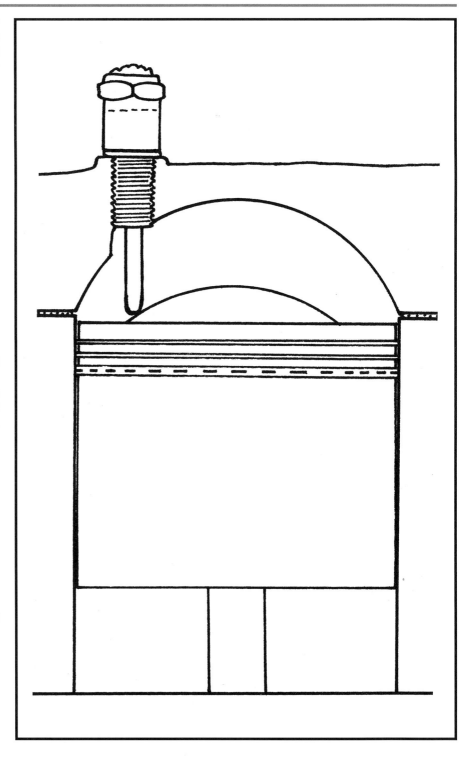

is rotated clockwise only and, when maximum upwards piston travel is first registered on the dial, crank rotation is immediately stopped: the position of the crankshaft damper in relation to the block-mounted pointer is then marked temporarily with a pen. The crankshaft is then turned (just a degree or two further) until the dial indicator needle moves again, showing that the piston is just starting to descend. The instant the dial indicator needle shows movement, crank rotation is stopped and the position of the crankshaft damper in relation to the block-mounted pointer marked temporarily with a pen. The top dead centre (TDC) mark on the damper is correct if it is in the middle of the two temporary marks you've just made. In other words, TDC is the middle of the period in which the piston dwells at the top of its stroke. It is the block-mounted pointer that is permanently adjusted if there are small changes to be made, not the damper TDC mark.

For assembled engines, the 'dead stop' device is more useful, and a very suitable device can be made using an old sparkplug. The insulator is removed from the spark plug and the electrode is ground off. A piece of 1/4 inch diameter rod which protrudes past the end of the thread is brazed into the centre of the sparkplug body. The length of the protruding rod will need to suit your engine.

If you can turn the engine over with the 'dead stop' fitted, the rod protrusion is too short. Caution. The engine must only be turned by hand, with all of the sparkplugs out. Start with the protrusion too long and shorten it progressively to suit the particular engine. The ideal stopping distance will be between 3 and 5 degrees before top dead centre (BTDC) when the engine is turned

IGNITION SYSTEM

Sufficient timing marks on damper rim.

clockwise and then anti-clockwise (**Caution!** - don't do this without the timing chain being correctly fitted).

Mark the position of the block mounted pointer in relation to the crankshaft damper with a pen at the point at which the piston is stopped in both directions of crank rotation. If the existing TDC mark is exactly in the middle between the two temorary pen marks, it is accurate. If not, the block-mounted pointer needs to be permanently adjusted to indicate true TDC.

Adding other timing marks

With true TDC having been determined as described, 12 and 14 degree BTDC marks are added to the existing 0 to 10 degree marks on the damper rim, as are 38, 40, 42 and 44 degree BTDC marks. You should also add appropriate camshaft timing marks.

The damper is usually marked while it is off the engine. A circle, the same diameter as the damper, is drawn accurately onto a piece of paper and marked with a TDC datum point and then (accurately) using a protractor with the other required degree markings. The damper is placed face down onto the piece of paper and the degree marks on the paper transposed onto the damper rim. The markings on the piece of paper are drawn on in reverse (anti-clockwise) so that they are on the clockwise side of the TDC mark on the damper as the marks are transferred from paper to damper.

As a point of interest, the C and D-Type engines of the 1950s had the ignition timing set when the engine was under test on one of their Heenan-Froude DPX 3.45 dynos, and not to any particular degree marking as such. With the engine at maximum torque, the distributor was set in the mean position that saw the maximum 'brake pull' registered on the dial of the dyno. Each engine was set to its own individual requirement, which effectively meant that there could be some slight variation in the setting of one engine to another (a degree or so). All of these engines were power curve tested up to, but never over, 5750rpm. This information was supplied to me by Harry Spears, a development engineer at Jaguar between 1952 and 1955,

SPEEDPRO SERIES

This damper has all of the necessary degree markings on it for ignition and camshaft timing (degrees vary for camshaft timing, depending on the cam specifications).

and on the Le Mans engines under the overall direction of Jack Emerson.

Each time one of the racing engines was rebuilt by Jaguar it was sent from the rebuilding workshop to the testing department to be 'run up' and tested. The ignition timing was always finally set at this stage and in this manner. Because of the frequency of racing engine maintenance it was never necessary for the ignition to be adjusted in the field. If the distributor's position was moved, the optimum as-tested setting would be lost, unless the distributor position had been marked beforehand.

STATIC (IDLE SPEED) ADVANCE - SETTING

The idle speed ignition timing requirements will vary. Most modified XK engines idle at between 1000 and 1200rpm. Use the highest amount of degrees of advance that allows the engine to attain the highest smooth idle speed. This setting will invariably be between 12 to 14 degrees before top dead centre (BTDC). It is acceptable for the idle advance to be as high as 16 degrees BTDC to get maximum idle speed smoothness. Try 16 degrees by all means to see if the engine speed increases and smoothness of running is maintained. However, what usually happens at 16 degrees is that the idle speed remains the same but the idle becomes rougher. These symptoms mean that you've gone too far. This amount of advance may work well if the engine has extra long camshaft duration (300 degrees plus). If your engine responds to 16 degrees of advance then use it, but do not set more than 16 degrees of idle advance for any petrol (gasoline) burning version of these engines.

If 16 degrees of static advance is used in an engine which has a 13 degree centrifugal advance distributor (26 crankshaft degrees) the engine will have a total advance of 42 degrees (excluding vacuum advance).

IGNITION ADVANCE - CHECKING

The ignition timing must advance centrifugally from the 12 or 14 degrees of static advance at idle to 'all in' total advance between 3000 and 3300rpm. This is quite easy to check with a strobe light (vacuum advance disconnected) if the crankshaft damper has been marked with suitable degree markings. Bear in mind that a lean air/fuel mixture will mean that an engine is likely to pink (ping) under acceleration, of course, and this must not be confused with the ignition timing being too far advanced.

Ignition advance is always measured in crankshaft degrees, never distributor degrees. This is because the easiest way of checking

IGNITION SYSTEM

the ignition timing is via the crankshaft damper after it has been clearly and accurately degree marked. Strobe light ignition timing is the only way to check and set the ignition timing of any high performance XK engine, simply because all of the variables are taken in to account in the reading. What you 'see' happening at the crankshaft damper in the light of the strobe is a completey accurate reflection of what the ignition system is doing at that particular engine speed.

IGNITION TIMING/TOTAL ADVANCE - FINDING OPTIMUM

The total amount of advance to use on these engines depends on several factors but, ultimately, the appropriate range of total advance is between 38 and 44 degrees. **Caution!** - 44 degrees is the definitely the maximum ever to be used on a high performance XK engine, and 38 degrees is generally the minimum.

A 13 degree distributor can be used to check and finalise the optimum amount of total advance, but it will mean temporarily setting the static advance to a higher setting than recommended. The multiples of spark advance using a 13 degree distributor (26 degrees at the crankshaft), involve setting the distributor with each of the following: 12 degrees of static advance (38 degrees of total advance); 14 degrees of static advance (40 degrees of total advance); 16 degrees of static advance (42 degrees of total advance); 18 degrees of static advance (44 degrees of total advance).

At 16 and 18 degrees of idle advance the idle might not be too good. However, for the short duration of the tests this is acceptable. To test the engine for the ideal amount of total advance within the range of 38 to 44 degrees before top dead centre (BTDC), start with 38 degrees and go up in two degree steps. Caution! - Try all four settings but do not persist if the 42 or 44 total advance settings cause the engine to pink (ping). Also, do not use full throttle under 3500rpm with the static timing set at either 16 or 18 degrees of advance (use two thirds throttle only). Care is needed here to test the engine for the amount of total advance that can safely be used.

Caution! - When testing any XK engine under load, DO NOT exceed 44 degrees of total advance. The engine could be damaged if it's run for a time with pre-ignition (pinking/pinging). Pre-ignition, resulting in serious engine damage, is quite possible when there is a combination of incorrect factors affecting in an engine. These factors are too much advance, too lean a fuel mixture, and too much engine compression for the octane rating of fuel being used. This is not to say that the XK engine is predisposed to pre-ignition (and consequent damage) if the slightest mistake is made when it's being tuned, rather a general warning that engine damage is possible if care is not taken.

Note that, with sufficient centrifugal advance mechanism spring tension, no additional (centrifugal) advance will be added to the static advance at idle speed. In most instances, unless the springs are too weak, centrifugal advance will not begin until beyond 1400rpm.

Total ignition advance requirements for a road going engine or for a racing engine are the same (disregarding vacuum advance which does not play a role in acceleration). The amounts of advance under discussion here are the amounts of advance required for a particular engine to develop as much useful power as possible over as wide a rev band as possible. Ignition advance is optimised to suit an individual engine and the modifications that have been made to it. When the total advance is right, the engine's mid-range and top end performance will be right too and this will be reflected in its measured performance. It is difficult to describe, but retarded engines have a 'heavy' sound to them and they are generally sluggish.

Engine testing to find the optimum ignition timing

You'll need the use of a decent length of road or track that can be used safely - not the public highway. What you are seeking is the smallest amount of advance that allows the engine to pull the maximum possible rpm in top gear. You'll need to repeat the test four times (total advance set, via static advance setting, at 38, then 40, then 42 and finally a maximum of 44 degrees BTDC) and record the results.

The other part of this test is that the total amount of advance has to be the setting that allows the engine to accelerate the quickest. This means, for example, that an engine could be set with 38 degrees of advance and pull maximum possible engine rpm in top gear. That same engine set with 40 degrees of total advance might, however, not pull any more revs in top gear but might allow the car to accelerate quicker. The acceleration rate factor being the vital difference. Check the rate of engine acceleration in the four total advance positions (38, 40, 42 and 44 degrees BTDC).

The two optimal amounts (for top gear rpm and acceleration) of total advance do not have to be the same, but they usually are on XK engines. Use the greater amount of advance as your ignition timing setting if they do prove to be different, but only if there is no pre-ignition (pinking/pinging) present.

SPEEDPRO SERIES

An alternative method of finding optimal ignition timing is to time the car's rate of acceleration between two speedo points in third or fourth gear using full throttle. Each run is made using the same engine rpm, the accelerator pedal is fully depressed the instant the first mark is passed and the car is accelerated to between 3000 and 5500rpm, for example. The time taken to cover the set distance can be noted using a stopwatch.

No gear changes are made, only full throttle is used, and the distance is the same each time. Any reduction, therefore, in the time taken to cover the distance, and any increase in rpm, means you are going faster. Conversely, any increase in the time taken to cover the set distance, and any reduction in terminal engine rpm, means you are going slower. Once practised and mastered, this is quite an accurate and simple test situation. Such tests, at the cost of a stopwatch, don't come much cheaper and the vehicle speeds used to achieve the data do not have to be excessive for public highways. Warning! - If you do use this method on public highways, due care must be taken to protect the safety of others and yourself.

Engines that have the correct amount of spark advance feel 'right' when accelerating through the gears. They have a sense of 'urgency' and, with an XK engine, that feel is generally unmistakable. Having the right amount of static advance at idle, the right rate of ignition advance and, finally, the right total amount of ignition advance is vital if an XK engine is to go as well as possible throughout the rev range.

Once the optimal total spark advance settings have been found, any engine being raced needs to have the timing adjustment checked before racing commences (on the day). Ignition timing has a habit of 'moving,' so you should always have a strobe light on hand when you're race prepping the car.

Summary

The following procedure will act as a guide to getting the ignition requirements of an XK engine more or less right first time.

Obtain a 13 degree Lucas distributor, remove the strong advance spring and replace it with a spring which is slightly weaker than the original spring still left in the distributor. Set the static ignition timing to 12 degrees before top dead centre (BTDC) and start the engine. Use a strobe light to see what the ignition advance is at 1200rpm (which could be the normal idle speed on some engines). The amount of advance should be 12 degrees at 1200rpm. If it's more than this, the replacement advance spring is probably too weak.

Gradually increase the engine revs and record at what rpm the ignition stops advancing: the engine speed should not be less than 3000rpm and not more than 3500rpm.

If necessary, alter the tension controlling the mechanical advance mechanism by substituting springs of different strength until the total advance is 'all in' between 3000 & 3500rpm. When this is achieved the engine is ready for road or track testing.

Note that because the engine has a 13 degree distributor fitted, and 12 degrees BTDC static advance, the maximum possible total advance is going to be 38 degrees before top dead centre (BTDC). Check what the exact amount of total advance is (should be 38 degree BTDC) and, if it is less, set it to exactly to 38 degrees.

For many engines the foregoing procedure will provide ideal ignition advance characteristics, for others it least a starting point has been provided. If you do need to make further adjustments, work carefully and make only one change at a time. Be prepared to spend time testing and, if necessary, altering the advance system/timing to achieve optimum settings. Too much power and overall engine efficiency can be missed out on if the ignition system is not dead right.

www.velocebooks.com/www.veloce.co.uk
All books in print • New books • Special offers • Gift vouchers • Forum

Chapter 6
SU carburettors

By far the most common carburettors fitted to XK engines are twin SUs. Twin 1 3/4 inch SUs were standard equipment on B-Type cylinder headed engines, such as those found in 3.4 and 3.8-litre Mark II and S-Type saloon cars, and twin 2 inch SUs were standard equipment on later XJ6 engines which had 3.4-litre and 4.2-litre engines (straight port cylinder heads). Twin SU HIF7s were also used on some later XK engines fitted with straight port cylinder heads.

The inlet manifolds of B-Type cylinder heads and straight port cylinder heads are not interchangeable, since the inlet port centres are different (this is reasonably common knowledge). 2 inch SUs can be fitted onto the B-Type inlet manifold if it is opened out to suit the larger carburettors. The two 1 3/4 inch diameter openings in the inlet manifold onto which the carburettors mount are opened up to 2 inches in diameter. This is easily achieved using a die grinder or a pistol drill fitted with a rotary file.

Plenty of modified engines go unbelievably well equipped with twin 1 3/4 inch SU carbs fitted with appropriate needles. In fact, retaining twin 1 3/4 inch SUs rarely results in a modified engine being under carburettored until well up into the rpm range. Larger carburettors, or more carburettors (triple instead of twin), are only needed if the engine has been sufficiently modified so that it can use more airflow. To say that swapping larger/more carburettors for the standard 1 3/4 inch twin SUs is a must for any modified engine is simply not true. It's just that there comes a point at higher revs where having more carburettors becomes an advantage. No modified road going engine realistically needs more than twin 1 3/4 inch or 2 inch SUs on either type of standard inlet manifold on either type of cylinder head. Racers will benefit from a two or three 2 inch SUs.

Triple 2 inch SUs were used on the standard 3.8 and 4.2-litre engines of E-Type and Mark X saloon cars. Both these engine capacities had the straight port cylinder head. In standard form triple carburettors offered some slight improvement in engine performance and represented the best of the factory fitted induction systems of the time: they looked more purposeful than twin SUs too! Secondhand triple 2 inch SUs are not cheap to buy these days, if you can find a set at all (one reason why sidedraught Weber or Dellorto carburettors often get fitted).

To sum up, twin 1 3/4 or 2 inch SUs are fine for road use, twin 2 inch SUS on B-Type manifolds or with an XJ6 head are good for racing. Triple 2 inch SUs - suitably jetted can be used for road or race.

Although standard twin carburettor inlet manifolds do not look particularly efficient, they're not actually too bad. They were originally

SPEEDPRO SERIES

designed to enable the close fitting of the carburettors to the XK engine since engine compartment space was limited (MkIIs and S-Types for example). These inlet manifolds are still very satisfactory for most uses but, on the basis of optimum airflow and air/fuel distribution, they're not as good as the triple carburettor manifolds. The 1961 and on twin 1 3/4 or 2 inch carburettor inlet manifolds, for example, are all reasonably well sized internally just as they come. The internal radiusing on the earlier B-type inlet manifolds, however, is not as good as it is on the later XJ6 inlet manifolds.

On both types of inlet manifold, the six short inlet manifold runners that feed the inlet ports are approximately 1.5in/38mm in diameter, which is a minimum of 0.125in larger than the actual inlet port of either type of cylinder head. The later 2in SU XJ6 inlet manifold is a better design than the earlier B-Type manifold in that the six short inlet manifold runners are slightly angled towards the nearest carburettor. The earlier B-Type inlet manifold's runners are not angled.

Fitting an XJ-6 straight port cylinder head (along with twin 2 inch SUs and appropriate inlet manifold) onto one of the older 3.4 or 3.8 litre engines is now fairly common. This has come about due to the now widespread availability of XJ-6 cylinder heads from scrapped or crashed XJ-6 cars. If XJ-6 cylinder heads are used on earlier blocks, the two rearmost waterway holes of later cylinder heads will need to be drilled, tapped, and then plugged, as they overlap the cylinder head gasket and block!

TWIN 1 3/4 INCH SU CARBURETTORS

Most standard 3.4 and 3.8-litre engines were equipped with twin 1 3/4 inch SUs fitted with a range of needles. The comprehensive factory range of needles for both of these engines, in order of approximate 'top end' leanness to richness, was DG, TM, TL, SJ, CI, TU, WO3, WO2, RF, RB, RC. All of these carburettors were equipped with red piston springs. Because of design constraints in 1 3/4 inch SUs, the needles are effective to approximately the 12th measuring point (every SU needle has a number of measuring points between shoulder and tip). This means that at full throttle and at maximum engine speed (carburettor piston at full upward travel), the 12th measuring point of the needle (or thereabouts) is metering the fuel.

For any of the engines equipped with twin 1 3/4 inch SUs, the TL needle is the baseline needle on the basis of good all round engine performance. Start with a TL needle and go richer as required. The factory fitted range of needles and their measuring point diameters are shown (lean to rich, left to right) in Table 1.

TWIN 2 INCH SU CARBURETTORS

Early 3.4-litre engines (early-1950s) and later 4.2-litre engines were equipped with different models of twin 2 inch SU carburettor and, consequently, both models of carburettor use different types of needle. The factory listed needle range included, in lean to rich order, NA, UM, UE for the later engines and VE,

	TL	TU	SJ	WO3	DG	WO2	RF	RB	RC
1st	0.099	0.099	0.099	0.100	0.100	0.100	0.100	0.099	0.099
2nd	0.095	0.095	0.095	0.095	0.098	0.095	0.095	0.095	0.0946
3rd	0.092	0.091	0.092	0.08775	0.096	0.091	0.090	0.0907	0.090
4th	0.089	0.088	0.0895	0.0845	0.0905	0.087	0.0863	0.0866	0.0855
5th	0.086	0.085	0.0875	0.0822	0.087	0.0835	0.0825	0.0825	0.081
6th	0.0835	0.083	0.0855	0.080	0.0836	0.081	0.0788	0.0784	0.0765
7th	0.081	0.081	0.0835	0.078	0.0804	0.0785	0.075	0.074	0.072
8th	0.0793	0.0793	0.0815	0.0755	0.0772	0.076	0.0712	0.070	0.0674
9th	0.0776	0.0776	0.0792	0.0735	0.074	0.0732	0.0675	0.0657	0.0627
10th	0.0759	0.0759	0.077	0.0712	0.071	0.071	0.0637	0.0615	0.0583
11th	0.0746	0.0746	0.075	0.069	0.0689	0.0683	0.060	0.0575	0.0537
12th	0.0733	0.0733	0.073	0.067	0.067	0.0657	0.057	0.0532	0.0492
13th	0.072	0.072	0.071	0.0653	0.0653	0.063	0.054	0.049	0.0446
14th	0.071	0.071	0.069	0.065	0.0636				

Table 1: Twin 1 3/4 inch SUs - needles.

SU CARBURETTORS

VR and 75 for the earlier engines (note that 75, VR and VE needles have 5/32 inch shanks). SU make U75, UVR and UVE needles which are the same as 75, VR and VE at the measuring points, but have different shank sizes (1/8 inch shank) for fitting into the later 2 inch SU carburettors. So, the complete range of needles used by Jaguar can be fitted to the later 2 inch SUs.

Of the needles mentioned, the NA can be ignored because it's too lean for high performance engines. The additional (non-factory offered) needle included in the listing is the UO. This needle offers more top end richness while maintaining factory calculated bottom end mixture strength. It's not possible go too rich with the mixture low down, as these engines just don't run correctly in this state. For twin 2 inch SUs the potential range of needles and their measuring point diameters are shown in Table 2.

The needles shown in the twin 2 inch SU tables are in lean to rich order, left to right across the page after the 11th measuring point of each needle. They are not all necessarily in order from the 1st measuring point to the 10th measuring point. The first three needles (UM, UE and UO), however, are completely in order. The VE is richer than all of the others between the 2nd and 6th measuring points and this is often just a bit too rich for some applications. 2 inch SUs use virtually the whole length of the needle to meter fuel, up to the 15th measuring point and beyond.

Spring loaded (swing) needle SUs

Later 3.4 and 4.2-litre twin carburettor Jaguar engines had 2 inch SUs which had 0.100 inch diameter main jets and spring loaded (swing) needles. The needles for these engines were BCX for the 3.4-litre engine and BAW for the 4.2-litre engines. These carburettors are quite suitable for use in any high performance engine. These two standard factory fit needles, shown in Table 3, are effective to the 15th measuring point.

For any high performance application of the 3.4 or 4.2-litre XK engines use BAW needles as a starting point and develop from there. Table 4 shows progressive alternatives to the BAW SU needles.

Other needles similar to BAM needles are BBP, BDA, BCQ. The problem with the SU carburettors, which have 0.100 inch spring loaded needles, is that for high performance applications there isn't the desired range of richer than standard needles. The BCE needle has the right amount of off idle and mid-range richness, but the top end's quite rich. The solution to this problem is to alter existing SU needles in the manner described in the SpeedPro book *How To Build & Power Tune SU Carburettors* by Des Hammill. This book covers the tuning of SU carburettors in great detail,

	UM	UE	UO	UVE VE	UVR VR	U75 75
1st	0.124	0.124	0.0124	0.124	0.124	0.125
2nd	0.1205	0.1205	0.1205	0.1178	0.119	0.119
3rd	0.1165	0.1155	0.1155	0.1125	0.1145	0.113
4th	0.114	0.1135	0.1135	0.108	0.1108	0.1087
5th	0.1123	0.1112	0.110	0.104	0.1075	0.105
6th	0.1104	0.109	0.108	0.1008	0.1038	0.1001
7th	0.1086	0.107	0.1055	0.0975	0.1004	0.097
8th	0.107	0.105	0.103	0.0943	0.097	0.093
9th	0.1056	0.103	0.101	0.091	0.0925	0.089
10th	0.1046	0.1015	0.0985	0.0877	0.088	0.085
11th	0.104	0.100	0.0965	0.0846	0.0835	0.081
12th	0.1032	0.099	0.094	0.0814	0.079	0.077
13th	0.1025	0.089	0.093	0.078	0.076	0.073
14th	0.1018	0.0965	0.0915	0.075	0.073	0.069
15th	0.101	0.095	0.090	0.0715	0.070	0.065
16th	0.1002	0.0935	0.089	0.0685	0.067	

Table 2: Twin 2 inch SUs - needles.

	BCX	BAW
1st	0.099	0.099
2nd	0.095	0.095
3rd	0.0924	0.0918
4th	0.0897	0.0887
5th	0.086	0.086
6th	0.0835	0.0827
7th	0.081	0.0799
8th	0.0785	0.0774
9th	0.0775	0.0755
10th	0.077	0.0735
11th	0.0765	0.0715
12th	0.076	0.070
13th	0.0755	0.069
14th	0.075	0.068
15th	0.0745	0.067
16th	0.074	0.065

Table 3: Twin 2 inch SU - (spring loaded) standard needles.

SPEEDPRO SERIES

	BAW	BAM	BCE
1st	0.099	0.099	0.099
2nd	0.095	0.095	0.095
3rd	0.0918	0.0915	0.090
4th	0.0887	0.088	0.0875
5th	0.086	0.0858	0.0835
6th	0.0827	0.0815	0.0795
7th	0.0799	0.0774	0.0768
8th	0.0774	0.0774	0.074
9th	0.0755	0.0742	0.0705
10th	0.0735	0.0694	0.0675
11th	0.0715	0.0673	0.0645
12th	0.070	0.0652	0.0615
13th	0.069	0.0672	0.0585
14th	0.068	0.065	0.0555
15th	0.067	0.063	0.0525
16th	0.065	0.061	0.0495

Table 4: Twin 2 inch SU - (spring loaded) alternative needles.

	BDY	BAM	BCE
1st	0.099	0.099	0.099
2nd	0.096	0.095	0.095
3rd	0.094	0.0915	0.090
4th	0.092	0.088	0.0872
5th	0.0897	0.0848	0.0835
6th	0.0855	0.0821	0.0795
7th	0.0824	0.0796	0.0768
8th	0.0801	0.0773	0.074
9th	0.0788	0.075	0.0705
10th	0.0778	0.073	0.0675
11th	0.0768	0.0713	0.0645
12th	0.0759	0.0692	0.0615
13th	0.0749	0.0672	0.058
14th	0.0738	0.065	0.0555
15th	0.0728	0.063	0.0525
16th	0.072	0.061	0.0495

Table 5: HIF7 twin 2 inch SU - (spring loaded) needles.

including reshaping needles. With correctly profiled and sized needles, these swing needle carburettors are as good as any other 2 inch SUs.

Altering existing needles, using the techniques suggested in the SU book, will produce needles which are ideally suited to your particular engine.

HIF7 SUs

The HIF7 SUs used on later XK engines also feature 0.100 inch spring loaded needles. They used BDY and BDX needles, of which the BDY is the richer needle. Using the BDY as the baseline needle and progress from there if necessary. In HIF7s, the needle is effective up to, approximately, the 14th measuring point. Table 5 lists needle types and measuring point sizes.

TRIPLE 2 INCH SU CARBURETTORS

The inlet manifold that Jaguar made for engines fitted with three 2 inch SUs was well designed and almost never requires any reworking to improve its efficiency. While these inlet manifolds do get modified for certain applications, most tuners should leave them well alone.

The UM needle, used on all E-Type and Mark X triple carburettor engines, is the baseline needle for high performance applications since it is the leanest of the range used as standard by Jaguar. The range of alternative needles to suit $1/8$ inch needle shank, 2 inch SU carburettors, is listed in Table 6 going progressively lean to rich from left to right.

The range of SU-made needles available is very comprehensive but, if the needles available do not work in your particular engine the solution is to modify needles, as described in the Veloce SpeedPro book *How To Build & Power Tune SU Carburettors* by Des Hammill. Needles can be altered slightly so that each needle measuring point size is exactly right for the particular engine.

All SU devotees should go to the trouble of learning how to alter SU needles to suit the exact requirements of their own engines. It really is worth doing.

	UM	UBUE	UO	UN	
1st	0.124	0.124	0.124	0.124	0.124
2nd	0.1205	0.1205	0.1205	0.1205	0.1205
3rd	0.1165	0.1165	0.1155	0.115	0.1165
4th	0.114	0.113	0.1135	0.1135	0.113
5th	0.1123	0.111	0.1112	0.110	0.110
6th	0.1104	0.109	0.109	0.108	0.107
7th	0.1086	0.107	0.107	0.1055	0.104
8th	0.107	0.1055	0.105	0.103	0.1005
9th	0.1056	0.104	0.103	0.101	0.0985
10th	0.1046	0.103	0.1015	0.0985	0.0965
11th	0.104	0.102	0.100	0.0965	0.0945
12th	0.1032	0.101	0.099	0.094	0.0915
13th	0.1025	0.100	0.098	0.093	0.0885
14th	0.1018	0.099	0.0965	0.0915	0.0855
15th	0.101	0.098	0.095	0.090	0.083
16th	0.1002	0.097	0.0935	0.089	0.0805

Table 6: Triple 2 inch SUs - needles.

SU CARBURETTORS

MODIFIED SU CARBURETTORS

For out and out performance, SUs can be modified to give much improved engine response but, unfortunately, at the cost of miles per gallon. Modifying SU carburettors really only involves leaving two things out, i.e. the oil in the dashpot and the piston spring. A small amount of oil is used for lubrication purposes, but that's all, and 'tailor made' needles have to be fitted. This all sounds reasonably simple and, basically, it is. However, the ideal needle profile has to be found through a progressive process, then the modified needles need to be identical carburettor to carburettor. Modifying needles is a complication, of course, but there is no other solution if ultimate acceleration performance is required from SUs. It's fair to say that a certain dexterity is required to make needles, but anyone who is handy with tools and who has the patience necessary will be able to master the technique with a little practice.

Once modified, SUs will use the same amount of fuel, or perhaps a bit more in some cases, as Weber or Dellortos tuned to give maximum power. The SpeedPro book, *How To Build and Power Tune SU Carburettors* by Des Hammill deals in detail with the modification of SUs and the information can be applied to twin or triple SU carburettor set-ups on any XK engine. You cannot buy the needles that are required for modified SUs and, consequently, standard originals will have to be altered to suit.

Note that while the throttle spindles can be modified by thinning them down, this procedure is not really necessary in the case of twin or triple 2 inch SUs as fitted to XK engines.

1 3/4 inch SU needles

Table 7 details dimensions (lean to rich from left to right) for modified needles for twin 1 3/4 inch SUs fitted to modified XK engines. The TL needle is the baseline or starting point needle for any XK engine equipped with twin 1 3/4 inch SUs. However, note that the TL needle will not work in a modified SU (without a piston spring or oil in the dashpot) but is included here to show the progression as the TL needle is modified. The measuring point sizes of the TL needles are very relevant from about the 8th or 9th measuring point onwards. Needles (ONE to SIX) can be fitted to these carburettors for testing without the piston spring fitted or any oil in the dashpot.

The recommendation is to make and try the needles in order (ONE to SEVEN) or you can follow the development process described in *How To Build and Power Tune SU Carburettors* and create needles to suit your engine's exact requirements. The needle data supplied in the table should be regarded as a guide, there will be some variation in optimum needle measuring point sizes depending on the particular engine's air/fuel requirements and general state of tune.

2 inch SU needles

The needle profiles detailed in Table 8 are approximations for use in XK engines that use two or three 2 inch SUs without piston springs and oil in the dashpots. Because of the individual

	TL	ONE	TWO	THREE	FOUR	FIVE	SIX	SEVEN
1st	0.099	0.099	0.099	0.099	0.099	0.099	0.099	0.099
2nd	0.095	0.089	0.089	0.089	0.089	0.088	0.088	0.088
3rd	0.092	0.086	0.086	0.086	0.086	0.085	0.085	0.085
4th	0.089	0.0845	0.084	0.083	0.083	0.0825	0.082	0.082
5th	0.086	0.083	0.0825	0.082	0.081	0.081	0.080	0.080
6th	0.0835	0.0815	0.081	0.081	0.079	0.079	0.078	0.078
7th	0.081	0.080	0.078	0.078	0.077	0.075	0.075	0.075
8th	0.0793	0.078	0.076	0.076	0.072	0.071	0.071	0.071
9th	0.0776	0.0765	0.0745	0.074	0.0695	0.068	0.067	0.067
10th	0.0759	0.075	0.073	0.071	0.069	0.065	0.064	0.063
11th	0.0746	0.0735	0.072	0.069	0.067	0.063	0.061	0.060
12th	0.0733	0.072	0.070	0.067	0.064	0.061	0.059	0.057
13th	0.072	0.071	0.068	0.065	0.061	0.058	0.056	0.054
14th	0.071	0.070	0.066	0.062	0.058	0.055	0.053	0.051

Table 7: Modified needles for 1 3/4 SU carburettors.

SPEEDPRO SERIES

	UM	ONE	TWO	THREE	FOUR	FIVE	SIX	SEVEN
1st	0.124	0.124	0.124	0.124	0.124	0.124	0.124	0.124
2nd	0.1205	0.117	0.117	0.117	0.117	0.116	0.116	0.116
3rd	0.1165	0.112	0.112	0.112	0.112	0.111	0.109	0.109
4th	0.114	0.110	0.110	0.110	0.110	0.109	0.107	0.107
5th	0.1123	0.109	0.108	0.108	0.108	0.107	0.105	0.105
6th	0.1104	0.108	0.107	0.107	0.106	0.105	0.103	0.103
7th	0.1086	0.107	0.105	0.105	0.1045	0.103	0.101	0.101
8th	0.107	0.106	0.104	0.1035	0.103	0.1015	0.099	0.099
9th	0.1056	0.105	0.103	0.103	0.100	0.100	0.097	0.097
10th	0.1046	0.104	0.102	0.1015	0.099	0.098	0.096	0.095
11th	0.104	0.103	0.1015	0.100	0.097	0.096	0.095	0.094
12th	0.1023	0.102	0.1005	0.099	0.095	0.094	0.093	0.092
13th	0.1025	0.101	0.0995	0.098	0.093	0.092	0.090	0.089
14th	0.1018	0.100	0.0985	0.097	0.092	0.090	0.088	0.086
15th	0.101	0.099	0.097	0.095	0.091	0.088	0.085	0.083
16th	0.1002	0.098	0.096	0.094	0.090	0.086	0.083	0.081

Table 8: 2 inch SUs (modified) - modified needles.

variation in states of engine tune, the dimensions given are not correct for each and every application, though they should put you in the ballpark. The final profiles of the optimum needles for your particular application may have to be arrived at after testing. The UM needle is the baseline or starting point needle.

These needles will not work without oil or piston springs in the carburettors. Because most triple SU engines have these needles fitted, this is the baseline size to start with. These original needles can, however, be altered to the dimensions as at ONE (Table 8). With needles of this shape the engine will run without oil and springs in the dashpot but it might not run all that well low down. ONE to SEVEN are guide sizes, the true dimensions are best found using calibrated rods.

The modified needle specification starts at ONE and goes to SEVEN (lean to rich from left to right). As can be seen, the 2nd, 3rd, 4th, 5th, 6th, 7th and 8th needle measuring points are where the major differences from the standard UM SU needle occur, and the first (ONE) needle for low engine speed response.

It's the difference in sizes between the 2nd and the 8th measuring points which enables the engine to accelerate better from lower down in the rev range. Smaller measuring point sizes provide a rich mixture which enables the engine to accelerate cleanly and without hesitation.

The recommendation is to make and try the needles in order (ONE to SEVEN) or you can follow the development process described in *How To Build and Power Tune SU Carburettors* and create needles to suit your engine's exact requirements. The needle data supplied in the table should be regarded as a guide, there will be some variation in optimum needle measuring point sizes depending on the particular engine's air/fuel requirements and general state of tune.

SU equipped engines which have had their needles modified to suit the exact requirements of the engine accelerate better than engines equipped with SUs which still have oil in their dashpots and piston springs fitted to them. Top end power will usually be better (though not always by much). The improvement in acceleration is the significant factor which is where SUs have always lagged behind similar Weber or Dellorto equipped XK engines.

CONCLUSION

Many people are of the view that XK engines fitted with twin 1 3/4 inch SUs just have to have grossly inferior engine performance to that same engine equipped with two or three 2 inch SUs. The factory itself, in fact, promoted the improved speed aspect of bigger and more is better. Experience has shown, however, that having twin 1 3/4 inch SUs on a 3.4 or 3.8 litre engine is no bad thing. All factory carburettor systems for these engines work well when the air fuel mixture ratios are absolutely correct for the particular engine.

In many instances, Jaguar owners are reluctant to lift the bonnets of their twin SU carburettored cars, especially when there are triple SU or triple Weber or Dellorto engines on display. Anyone running twin SUs should forget about looks and concentrate instead on making sure that the SUs on their engine are jetted correctly, and that the true level of possible performance is being achieved. Twin SUs of either size can be used on modified engines for road use with complete confidence, though most people tend to want to use 2 inch SUs. For racing use, 2 inch SUs will usually, though not always, offer improved absolute top end performance. Triple 2 inch SUs, or triple Webers or Dellortos, are suitable for road or track use.

One thing about carburettors and inlet manifolds is, of course, that it does not take all that much to change from one sort to another. This means that the original type of carburettion system can be run first, and fully developed, and then an alternative system tried and tested. With an engine fully tried and tested with twin SUs fitted to it, base line figures will be known and direct comparisons can be made. The difference may not be as much as you would imagine! Twin SUs are as simple as it gets on XK engines and are the logical starting point.

After running correctly sorted out twin SUs on an XK engine, and then going to all the expense and trouble of changing to triple SUs or triple Webers or Dellortos, many people have asked themselves afterwards whether it was at all worthwhile.

Chapter 7
Weber & Dellorto carburettors

The most important thing to note about relatively exotic induction systems like triple sidedraught Webers and Dellortos, is that there will be no real increase in power until the engine is modified internally. If triple Weber carburettors are fitted to a standard engine the results will often be disappointing. In most instances, the fuel economy will drop off quite dramatically too. Also, if these particular carburettors are not set up correctly, the fuel economy could be half what it should be, with no gain, or even a loss, in power.

You can't just buy a set of triple sidedraught carburettors and bolt them to an XK engine and expect miracles. Tuning XK Jaguar engines is just not like that. The fact that there always seem to be highly priced, triple sidedraught induction systems on the second-hand market seems to prove the point. These induction systems usually being sold because they did not work well, and getting rid of them is seen as the only solution to getting some money back. Few people will sell these induction systems because they were fantastic on the engine! Most people just can't get them tuned correctly and give up.

The first thing to do when buying a Weber or Dellorto (either make will give similar performance) set-up is to check the jetting and compare it to the suggested jetting listings in this book. You should, of course, also check the overall condition of the carburettors, especially Dellorto idle adjustment screws with fine threads.

In almost all circumstances the jetting installed in these carburettors will be wildly out. Jetting that differs from what is suggested in this book will have to be altered. This will be expensive, as will be repairing the carburettors if they are damaged. Very often the cost of rejetting and repair can be knocked off the purchase price. Irrespective of what is wrong with a sidedraught set-up, expect to pay substantial money for it. There are not many engine parts that are so sought after when secondhand.

Fitting Weber or Dellorto carburettors to XK engines has a dramatic visual impact. Further, a correctly modified engine, with these instruments correctly tuned, will not only look good, but have the sparkling performance to match. However, for every XK engine that goes very well when fitted with these carburettors, there are probably five that don't ...

If your engine is equipped with Weber or Dellorto carburettors and lacks performance, don't despair - whatever's wrong can be fixed. The jetting can be altered quite easily by merely substituting parts. Complications only arise when the carburettors are damaged and parts cannot be changed.

The jetting specifications listed in this book are realistic, tried, tested and proven for average high performance applications where mainly standard

WEBER & DELLORTO CARBURETTORS

engine parts are going to be used and where the revs will not exceed 6000rpm. While larger chokes and their associated jetting are sometimes used on XK engines, the applications are likely to be highly tuned units used in all-out competition. Such engines, often viewed with a degree of 'expendability,' are usually rigorously maintained on a very frequent basis to ensure reliability.

Jaguar fitted sidedraught Weber carburettors to its Le Mans racing engines in the early 1950s to improve the engine performance over that obtained with SU carburettors. Harry Spears fitted the first set of Weber carburettors to an XK engine, ran it, re-jetted the carburettors a couple of times, and clearly remembers that the top end power was eventually improved by 20bhp over the SUs. SU's Peter Knight spend hours trying to improve the top end performance of the SUs, but it couldn't be done, and sidedraught Webers (fitted with 38mm chokes) were universally adopted by Jaguar for its Le Mans racing engines.

CHOKE SIZES & JETTING

There are basically two choke size options for the three engines under discussion here - 38 or 40mm. If the engine you have is road going, consider 38mm chokes to be ideal. If the engine you have is for racing purposes, consider 40mm chokes to be ideal. In some instances an XK engine may well go better on the road with 36mm chokes and associated jetting: it just depends on how the car is driven. For lower rpm use, mainly between 1000 and 5000rpm, all three engine sizes are likely to respond well to 36mm chokes and associated jetting.

If the choke size is taken down any lower than 36mm, the top end power will usually start to drop off, although engine acceleration from idle up to about 4500rpm will often be improved. Fitting small chokes almost always means good acceleration, though maximum power will be reduced.

Fitting very large chokes (42 or 43mm chokes in 48mm carburettors, for example) very often results in the best maximum power being produced at very high rpm (above 6000rpm) but a much less tractable engine below this. Having such large chokes and such large carburettors is not recommended for these engines. It is a known fact that for the sake of reliabilty the majority of these engines are not going to be revved more than 6000rpm, even in a racing situation, and, consequently, 40mm chokes are about right for just about all racing applications.

Experimenting with 41 and 39mm chokes and associated jetting on a racing engine may lead to an improvement in engine efficiency over 40mm chokes, but this is not usually the case. Experimenting with 37 and 39mm chokes and associated jetting on a road going engine may also lead to a slight improvement in 'driveability.' How well an engine responds to larger or smaller chokes and associated jetting will depend on how well it's modified. Making even slight changes in jetting is quite expensive with Weber and Dellorto carburettors, since new parts (times six) are not cheap. It's my experience that 38mm chokes and associated jetting seem to be best.

It goes without saying, of course, that any engine will have to be jetted correctly. The initial cost of doing this with Webers or Dellortos will be reasonably substantial. If sensible choices, based on known parameters, are made in the first place the optimum jetting for the individual application will found quite quickly and at minimum cost.

A common reason why sidedraught Weber or Dellorto carbs fail to produce good overall engine power when fitted to an XK unit is that the engine does not have enough ignition advance. Replacing the distributor with one that has suitable static advance, total advance, plus a quicker rate of advance than standard, can be the solution. The basic problem may be ignition rather than carburation.

Another book in the SpeedPro series, *How To Build And Power Tune Weber And Dellorto DCOE & DHLA Carburettors (second edition)* by Des Hammill deals in depth with sidedraught Weber and Dellorto carburettors and optimising their performance for an individual application.

Each engine will have to be fine tuned, of course, with slight adjustments to jetting being made as necessary and all settings optimised. However, starting point jetting for any XK engine is as follows -

DCOE Weber (38mm chokes)
4.5 auxiliary venturi (small central hole)
150 - 155 mains
F2 emulsion tubes
170 - 200 air correctors
45 pump jets
45 F9 idle jets
7.5mm fuel shut off height
15mm full droop setting

DCOE Weber (40mm chokes)
4.5 auxiliary venturi (small central hole)
160 - 165 mains
F2 emulsion tubes
170 - 220 air correctors
45 pump jets
45 F9 idle jets
7.5mm fuel shut off height
15mm full droop setting

SPEEDPRO SERIES

Dellorto DHLA (38mm chokes)
155 - 163 mains
180 - 200 air correctors
8011.1 auxiliary venturi
7777.6 emulsion tubes
60 accelerator pump jets
62 - 65 idle jets
7850.1 idle jet holder
15mm fuel shut off height
25mm full droop setting

Dellorto DHLA (40mm chokes)
160 - 168 mains
180 - 200 air correctors
8011.1 auxiliary venturi
7777.5 emulsion tubes
65 accelerator pump jets
65 - 68 idle jets
7850.1 idle jet holder
15mm fuel shut off height
25mm full droop setting

TUNING CARBURETTORS
Before starting the engine

With fine thread idle mixture adjustment screws, as found on Dellorto carburettors, turn each screw out 6.5 turns from the fully wound in (but only lightly seated) position.

For Weber carburettors, turn each of the idle mixture adjusting screws out at least 1.5 turns from the fully turned in (but only lightly seated) position.

Balancing airflow

Apply sufficient throttle to the carburettors so that the engine will at least start. Once the engine is running (however roughly) and warmed up, equalise the airflows of each of the three carburettors using the idle speed adjusting screw on each carburettor in conjunction with a flowmeter (Gunson, or similar). Expect a modified engine to idle at not less than 1000rpm and maybe up to 1200rpm. Individual carburettor airflow is adjusted, via the individual idle screws, with the three carburettor throttle linkages undone.

Once idle airflow is matched, the linkage clamps are tightened very carefully- so as not to disturb or alter the individual carburettor throttle arm settings - and the accelerator pedal pressed slightly to open the throttles. The airflow through each carburettor is then checked to see if it is still identical. Note that the accelerator pedal is used, as opposed to pushing the linkage inside the engine compartment, so that all possible wear and looseness and 'slop' in the linkage is removed from the action.

No triple carburettor engine will ever perform well if the carburettors are not synchronised perfectly (the pairs of cylinders will not receive equal amounts of air).

The object of the airflow balancing exercise is to make sure that the carburettors have the same amount of air going through them at idle and when the throttle pedal is pressed. Once this has been done, check that the throttle butterflies really are fully open when the accelerator pedal is hard down. Failure of the throttle mechanism to open the butterflies fully is one of the commonest reasons why many engines do not go as well as they could. You should check the throttle butterflies for full opening before any race day.

Pump jets & inlet/discharge valves

Although unlikely, there are some instances when 40 pump jets might well be enough on Weber carburettors. Some engines, on the other hand might well need to be fitted with 50 pump jets. The pump jet size is, to a small degree, dependant on the accelerator pump inlet/discharge valve size. Some inlet/discharge valves, which come in various sizes, are not discharge valves at all, they are inlet only. These are easy to recognise as they do not have a small hole drilled in the side. A zero valve has no hole in the side of it and the whole shot of fuel will be pumped into the engine. In these instances a smaller pump jet might well be used successfully. Consider the discharge aspect of the accelerator pump inlet valve as a tuning device.

Increasing the drilling size in the side of the valve will tend to reduce the amount of fuel pumped into the engine when the accelerator pedal is depressed. Conversely, reducing the drilling will tend to increase the amount of fuel pumped into the engine. There are limits as to how effective this is but, nevertheless, it is a way of increasing or reducing the pump shot size for a given pump jet size.

Having the minimum amount of fuel pumped into the engine (Dellorto or Weber equipped) each time the engine is accelerated, is the ideal situation. Insufficient fuel pumped into the engine when the accelerator pedal is pressed causes hesitation. Too much fuel pumped into the engine when the accelerator pedal is pressed leads to a lot of black smoke from the exhaust pipes and a degree of engine response sluggishness. When an XK engine is set up correctly there is no hesitation at all during engine acceleration, and only a slight puff of black smoke (unburnt fuel) from the exhaust pipes. It's just a question of getting the amount of fuel pumped into to the engine just right.

Air correctors

The air corrector size, which alters the top end mixture, selection is finalised through testing. For instance, if the engine has a slight miss at 6000rpm, reducing the size of the air corrector from 200 to 190 or 180, for example, will usually cure the problem (if the problem really is a carburation one, that is). Conversely, if the engine is

WEBER & DELLORTO CARBURETTORS

running well through to 6000rpm with 180 air correctors fitted, changing the air corrector to a 190, 200, 210 or even a 220, may result in a slight improvement in top end engine performance. If the engine starts to miss slightly when tested at high rpm (i.e. the mixture is too lean) the air corrector is too large: go back to the previous size that did not cause the engine to miss.

For those who want to use the original 45mm DCO3 type side draught Weber carburettors, all parts are available from Norman Seeney Ltd. (Tel: +44 (0) 1527 892650, Fax: +44 (0) 1527 893017).

Triple DCO3s on this D-Type Jaguar engine.

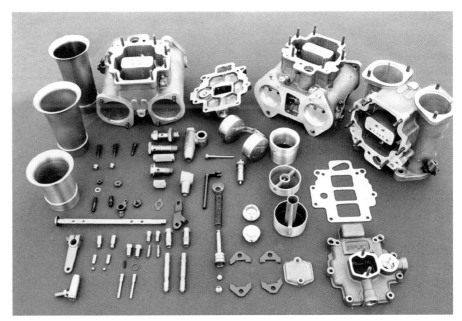

Just some of the components available from Norman Seeney Ltd.

Chapter 8
Exhaust system

Although the standard cast iron three-into-one exhaust manifolds are not particularly restrictive, they can be improved upon for high performance applications. The improvement is to fabricate or buy tubular exhaust manifolds. Since there are not many 'off the shelf' tubular exhaust manifolds available for XK engine installations, in most cases exhaust manifolds have to be custom made.

Many workshops specialise in making exhaust systems, for any application, and making one for an XK engine does not normally present any problems for experienced engineers. Firstly, though, two 0.25in/6.35mm thick cylinder head manifold mounting plates will have to be cut out of mild steel flat bar or plate. These two plates can be marked out using the standard exhaust manifold gaskets as templates.

The XK exhaust system is divided between the front three cylinders and the rear three cylinders. Each set of three cylinders exhausts into three primary pipes, then a main pipe and finally to atmosphere.

PRIMARY PIPE LENGTH

There are two basic designs that can be used as alternatives to the standard exhaust manifolds. The main difference between the two systems concerns the overall length of the six primary pipes. The primary pipes can be 'short' or 'long.' There's not much difference in effect with either system so, in broad terms, use the design that suits your particular engine installation best.

If the primary pipes are going to be short they'll be between 18 to 22 inches/457-559mm long. In an ideal situation, all of the primary pipes will be of equal length. This will not always be possible, however, because of space limitations in some engine installations. Fortunately, there is little measurable power loss on XK engines when the primary pipes are an inch or two different in length.

The short primary pipe exhaust manifolds are almost always the easiest to make up and fit into any installation. By comparison, the long primary pipe exhaust manifolds can be an absolute nightmare to make and fit. A long primary piped exhaust system will have primary pipes of 33 to 36 inches/838-914mm in length.

If an exhaust manifold is to be made up to fit a MkII saloon car, for instance, a short primary pipe exhaust manifold is the easiest to make and fit. Another important point, of course, is that the engine has to be maintained and the exhaust systems will occasionally have to be removed. It can get pretty annoying having to refit a complicated exhaust manifold. Simplicity is the best policy when it comes to exhaust manifolds, and servicing should be a major consideration.

The main advantage of the short primary pipe manifold is the fact that the three primary pipes join into one main pipe before they turn to go

EXHAUST SYSTEM

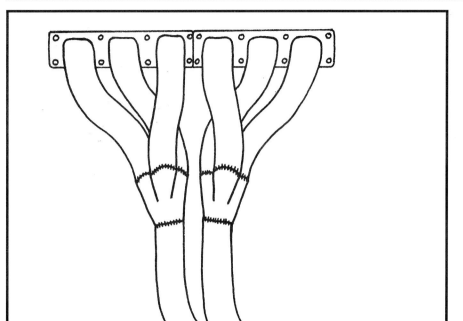

Short primary pipe-type exhaust manifold.

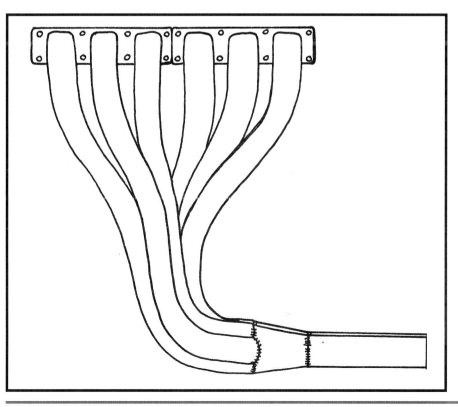

Long primary pipe-type exhaust manifold.

underneath the car. If a long primary piped exhaust manifold is going to be used, all six primary pipes go down and turn before joining their respective main pipes.

PRIMARY PIPE DIAMETER

There are three primary exhaust pipe diameters to use on Jaguar XK engines: 1³/₄ inch, 1⁷/₈ inch and 2 inch (45, 48 & 51mm) outside diameter. Although all three are acceptable, the larger the primary pipe diameter, the higher the point at which peak torque occurs in the rev range. When you look at an engine's power curve, the point of maximum torque can be clearly seen. Fitting larger diameter primary pipes will raise the rpm point where maximum torque occurs, though not the amount of torque actually produced.

Because the torque curve has been moved up the rev range, the engine will have more torque (and a greater sense of urgency) at higher rpm than at lower rpm. This will be quite noticeable in racing situations because, instead of reaching a peak and then losing urgency as the engine approaches maximum rpm, there will be no such loss since peak torque is now closer to maximum rpm. However, there are limits to how big the primary pipe diameters can be.

Having 1³/₄ inch diameter primary pipes on any of these engines which will usually work between 3000 and 6000rpm is about right. By all means go larger if you prefer, but you may not get any noticeable improvement in power for your trouble. You have to bear in mind, of course, that the XK engine produces good torque over a very wide rpm band anyway.

The most commonly used primary pipe diameters are either 1³/₄ inches or 1⁷/₈ inches. In relation to both these diameters the cylinder head exhaust port is larger and is of rectangular shape, however either pipe can be reshaped to fit. In each case the pipes will have to be cut, flared, reshaped to rectangular, and have an extra piece of material (triangular in shape) welded into them. This extra piece of material is always fitted on the underside of the pipe so that it can't be seen.

Although 2 inch diameter primary pipe is big enough to fit the port without being flared, it will still require reshaping. Once a 2 inch diameter pipe is reshaped to match the rectangular exhaust port, it can simply be welded onto the header plate. The trouble with a 2 inch primary pipe is that it's a bit large for engines that will normally operate between 3000rpm and 6000rpm. Also, in some instances, having six pipes of such large diameter in a confined space makes engine servicing more difficult.

For engines limited to 6000rpm, there will be little advantage (if any) in fitting primary pipes larger than 1³/₄ inches. The larger diameter pipes will only bring a worthwhile advantage to all out racing engines turning high rpm on a continuous basis.

MAIN PIPE DIAMETER

The three primary pipes then pass into a trifurcated joint which takes them into the main pipe. This main pipe is usually 2, 2¹/₈ or 2¹/₄ inches (51, 54 or 57mm) in diameter. 2 inch diameter main pipes are quite sufficient for any of these engines. Many engines do, of course, sport the larger pipe sizes mentioned here and go very well with them, this is because there is a reasonable amount of latitude with exhaust system pipe diameters on these engines.

Smaller exhaust pipe sizes than those mentioned can cause restriction problems and power loss, especially as the revs rise above 5000rpm. High performance XK engines have been fitted with tubular exhaust systems with 1⁵/₈ inch (41mm) diameter primary pipes and 1⁷/₈ inch (48mm) diameter main pipes. The engine response with such systems fitted is excellent for lower rpm use (1000 to 5000rpm). Exhaust systems with pipe sizes like these are excellent, and in fact ideal, for road going use where the engine will almost never be taken above 5000rpm.

SILENCERS (MUFFLERS)

The XK engine is quite difficult to keep quiet whilst at the same time maintaining efficiency. Although fitting two straight-through absorption type silencers (one for each main pipe) will reduce the noise, they'll never produce street legal noise levels. On some cars, however, there simply isn't very much room to fit anything else.

The only way to quieten one of these engines for road use is to fit baffle type silencers, of the type fitted by Jaguar as standard equipment. The problem with most original equipment standard replacement silencers is that they are quite expensive and yet still corrode quickly. Cheaper alternatives do exist, thankfully, and are termed 'universal fitting.' The problem with these universal fittings is finding a suitably large bore muffler that will fit the available space.

Another consideration (besides space) when fitting a non-standard muffler is to make sure that its main pipe diameter is as large, or larger, than the diameter of the main pipe coming from the manifold. The muffler's main pipe diameter needs to be a minimum of 2 inches. The usual universal fitting silencer is an oval wrapped muffler with one central pipe and one offset pipe. Check if the mufflers are flow directional (look for an arrow). Most are not, but some are ... Many of these

EXHAUST SYSTEM

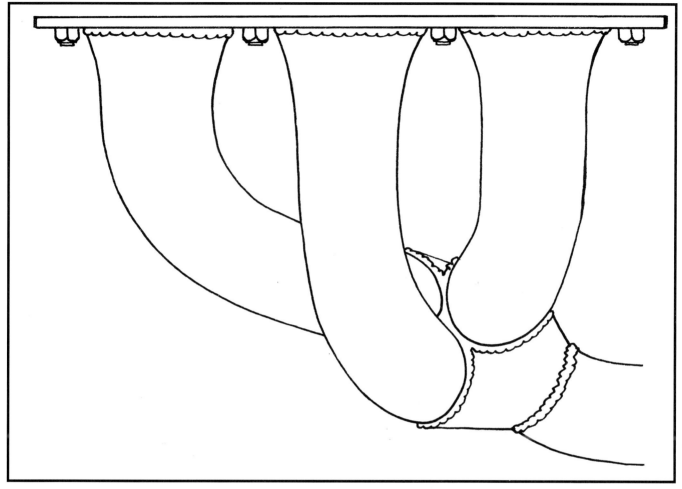

A 1 3/4 inch diameter primary pipe has to be flared to suit the cylinder head apertures. This takes a lot of work, but looks neat when finished.

universal fitting silencers are quite compact, and much more efficient than they are given credit for and are very reasonably priced.

It's fair to point out that, in many instances, these universal fitting silencers are made to a price which is well below that of a similar standard equipment. Since quietness is always a top priority for the companies that make these silencers, it's fair to say that such products could be slightly more restrictive than standard. However, as XK engines effectively have two separate exhaust systems, it is unlikely that silencers which are a little more restrictive than they need to be will cause significant power loss.

Another alternative is to look for a silencer type from another vehicle that has the required pipe size and is within the physical size dimensions required for the individual application. In many instances, exhaust specialists have a very large range of silencers from which to choose. Going to an exhaust specialist and having a good look to see what's are available 'off the shelf' is the first step, once you know the dimensions required.

Two silencers can often be fitted side by side in the underbody space usually occupied by one large silencer. Sometimes, oval type silencers can be turned 90 degrees to fit an installation. There is a considerable degree of latitude possible with silencers.

Chapter 9
General information

CAMSHAFTS

Before starting to assemble the cylinder head check the camshafts for freedom of rotation. This means fitting the new bearings into the cylinder head and into the camshaft bearing caps, and then installing the camshafts without the valves and valve springs in place. When the caps are all fitted and torqued up, the camshaft is rotated by hand. There must be no binding whatsoever. Binding can be caused by a bent camshaft, a damaged camshaft journal, or a problem with the bearing shells. **Caution!** - Do not proceed with the cylinder head assembly until the problem is found and resolved.

TAPPET ADJUSTMENT/VALVE STRETCH

When the engine is first assembled, make sure that the tappet clearances are exactly right (no less than 0.010in/0.254mm, no more than 0.011in/0.279mm). On a racing engine, it's vitally important to check the tappet clearances frequently to make sure that they do not close up. If the clearances do close up the engine may well become hard to start because the valves are not closed for the required duration: if anything, it's better that valve clearances are slightly too loose than too tight.

Apart from slight reductions in valve clearances caused by the engine 'settling in'/'bedding down' it's also possible that the valve stems are getting longer (stretching).

Caution! - Valve stretch does happen from time to time and such valves must be replaced: if a valve stretching problem is left unchecked, the valve head can actually fall off and cause serious damage to the engine. If a tappet clearance tightens up from its original 0.010in/0.254mm to as little as 0.005in/0.127mm, for example, reset the clearance and run the engine again for about the same interval. Check the tappet clearance again and, if it has tightened up by a similar amount again, replace the valve. This is a very simple way of checking to see what is happening with the valves without taking the engine to pieces.

ENGINE BALANCE

XK engines were all very well balanced when they were originally made. However, because of the age of these engines now, and the possibility that original parts have been changed over the years (as well as for complete piece of mind), it's advisable to have your high performance engine rebalanced.

This means having each piston, its gudgeon (piston/wrist) pin, circlips and piston rings collectively weighed. Each piston combination must be within 1 gram of the other 5. The weight of each end of each connecting rod, too, must be checked (bearings included), and end-for-end weights must all be within 1 gram of each other). Although this sounds very exacting, to people who rebalance engines it's all in a day's work.

GENERAL INFORMATION

The crankshaft should be dynamically balanced on its own. The damper, the flywheel and the clutch pressure plate are then added one at a time so that they can be individually balanced. The flywheel must be located accurately on the back of the crankshaft. There must be no possibility of the flywheel settling in a different position. The dowels must be a tight fit in the crankshaft flange, and the flywheel must be a tight fit onto the dowels. The flywheel needs to be marked for one position fitting.

Be aware that just because a pressure plate is advertised as being "pre-balanced" it does not mean that it will be when fitted to your flywheel. Whilst it's perfectly true that the manufacturers have balanced the pressure plate on their equipment, your flywheel dowels may be in a slightly different position, for example, making the whole manufacturer's exercise irrelevant. Rebalancing the pressure plate on your own flywheel will prevent any possible problems.

Caution! - Always fit a new diaphragm clutch pressure plate when the engine is rebalanced: if the pressure plate fails at high revs, the engine's balance will suddenly be lost and the chances are that the crankshaft will break too. Crankshafts break in the rearmost web of the crankshaft so, while it may appear that just the pressure plate failed and fitting a new one will cure the problem, this is rarely the case and there may well be a crack in the crank. You have to remember that these are long stroke engines with quite a lot of weight spinning around. At 6000rpm they will not tolerate gross imbalance at all. If a pressure plate fails, replace the crankshaft.

The crankshaft damper rubber deteriorates over time too (bear in mind that in some cases the rubber is 40 or 50 years old). This can mean that the damper is defective, although you might not be able to tell just by looking at it. Damper rims on old dampers have also been known to come off. The solution is to fit a damper from a newer lower mileage engine.

CRANKSHAFT THRUST WASHERS & CLUTCH PRESSURE PLATES

One part of the XK engine's design that is not as good as it might have been concerns the rear facing thrust washer on the crankshaft. When the clutch is depressed, the crankshaft is pushed towards the front of the engine and against the rearmost and rear facing thrust washer. Although this isn't normally a problem on a standard engine, it can be a major problem when a heavy duty pressure plate has been fitted. Some uprated pressure plates are very strong. The pressure plate resistance is then so strong that it will force the crankshaft's thrust surface against the thrust washer's surface which will then wear quite rapidly.

There are a few things that can be done to reduce this problem to an acceptable level. The first is to make sure that the crankshaft's rearmost thrust surface is completely flat, 90 degrees to the crankshaft axis, and has a mirror finish (like a crankshaft journal). If the thrust surface of the crankshaft on this side is blemished in any way you can expect problems. Only a perfect crankshaft thrust surface will do, especially if an uprated pressure plate has been fitted.

If the thrust washer surface of the crankshaft is scored or blemished it can be reground by an engine reconditioner/engine machine shop. Before getting them to regrind your crankshaft, though, explain to them that the surface must be absolutely flat and smooth and have a mirror finish. It's not easy to regrind the thrust surfaces of a crankshaft to such a high standard because the side of the grinding wheel is used. Polishing the surface also takes time and patience to get right. **Caution!** - In this instance, though, nothing less than perfect will do. Pay whatever it takes to get it right, and do not accept less.

Make sure that the crankshaft endfloat clearance is as near to 0.003in/0.076mm as possible. 0.004in/0.101mm is okay but more than this is too much. Anything above 0.006in/0.152mm is definitely too much.

Caution! - When replacing the rearmost thrust washer, only fit genuine Jaguar components or heavy duty aftermarket ones (copper in colour): never fit plain white metal thrust washers (silver in colour). The foremost thrust washer, however, can be a white metal replacement as there is no clutch load on it.

Caution! - Always check and, if necessary, replace the rearmost thrust washer whenever the engine is dismantled for a rebuild. Replace it if there is any sign of wear (this can be measured by checking the endfloat of the crankshaft with a feeler gauge).

Caution! - To help thrust washers last as long as possible, only fit the weakest up-rated pressure plate which is sufficient to prevent clutch slippage. For competition use, 2200 to 2600 pounds of clamping pressure should be ideal. Another way to prolong thrust washer life is to only depress the clutch pedal when you are going to move off or change gear: don't 'ride' the clutch at all. With a strong pressure plate fitted (3000 pounds of clamp or more), the less often the clutch is used the longer it and the thrust washers will last. Finally, set the clutch travel so that the pedal will only depress enough to disengage the clutch. This may mean fitting an adjustable pedal stop.

SPEEDPRO SERIES

As you can see there are quite a few things that can be done to prevent premature thrust washer wear. However, once a much stronger than normal pressure plate is fitted to the engine, thrust washer wear is inevitable no matter what is done. Adopting a slightly different driving style, though, can pay dividends.

AP Racing make uprated coil spring pressure plates under part number CP2413-1 for the 2600 pound clamping pressure one and CP2413-2 for the 3270 pound one. Most applications will be able to use the CP2413-1 pressure plate in conjunction with the four paddle cerametalic driven plate CP2516-2.

Caution! - While there is no doubt that the CP2413-2 pressure plate is very firm and will cope with huge torque, if at all possible, avoid using it as it's likely to cause high thrust washer wear. If you do decide to use this pressure plate, limiting the disengagement is quite feasible especially when used in conjunction with a hardwearing cerametalic driven plate (adjustment is more or less always maintained).

9.5 inch standard clutches came with 1600 pounds of clamping pressure. Uprated units are available from AP Racing with 1900 pounds of clamp (part number CP2345-8) or 2400 pounds of clamp (part number CP2394-14). CP2394-46 provides 3600 pounds of clamp pressure, but expect thrust washer problems with this unit. For 10 inch clutches part number CP2789-1 gives 1900 pounds of clamp pressure and CP2789-4 at 2200 pounds. These pressure plates are used in conjunction with CP2495 cerametalic driven plates.

Identifying which clutch size is relevant to your engine is essential. Consult the AP Racing catalogue if you have any doubts or AP Racing direct by e-mailing them at sales@apracing.co.uk for details.

Summary

Standard original equipment replacement pressure plates and driven plates are suitable for most road going high performance applications.

For racing purposes, a cerametalic driven plate is required; a rigid one being better than a sprung one in most instances. Keep the clamping pressure within the 2200 to 2600 pound range if possible. **Caution!** - A new original equipment pressure plate/cover assembly is not strong enough to use with a cerametalic driven plate.

If you have a cerametalic paddle clutch specially made for you, be aware of the fact that all Jaguar pressure plates are designed to work with driven plates that are 8.4mm thick. The pressure plate/cover assembly relationship must be maintained if the original specification clamp pressure is to be realised.

CLUTCH (DRIVEN) PLATES

The standard Jaguar clutch (driven) plate is pretty strong. For all road going use, even with a modified engine, they'll almost always prove to be adequate.

Problems with the clutch plate only arise when an engine is used in competition (because of the serious abuse that the clutch is likely to get). The solution for motorsport is to fit a cera-metallic lined 'paddle' type clutch plate. The advantage with these is that they do not wear much, even when abused to ridiculous levels. Because it's actually quite difficult to wear out one of these clutch plates, they are well worth the money (compared to how many normal woven type clutches you would have to fit and replace to equal them). Essentially, it's a case of 'fit it and forget it.'

Uprated pressure plates are recommended for use with a cerametallic lined 'paddle' clutch plate. However, if a very strong pressure plate is fitted the thrust washer problem will come into play. The solution to this problem is to try a genuine new standard pressure plate in the first instance and see whether or not it is sufficient for your particular application. If the clutch slips, the pressure plate will have to be up-rated.

Caution! - Because cera-metallic 'paddle' type clutch plates are often special orders made up to suit a particular application, there is room for clamping pressure error. This error, if present, will cause clutch slippage. The new cera-metallic clutch plate must be of the same thickness as a standard new clutch plate at the very least (or up to 0.010in/0.254mm thicker). If the cera-metallic clutch plate is thinner, the clamping pressure will not be the same and the clutch may slip. The more modified the engine the more chance of clutch slippage. The time to get all of this right, of course, is when the engine is out of the car!

Measure a new standard clutch plate designed to fit your model of Jaguar and make sure that the cerametallic clutch plate 'pucks' are the same thickness (minimum), or up to 0.010in/0.254mm thicker. If the 'pucks' are thinner than the thickness of a new standard clutch plate, reject the cera-metallic plate or machine the flywheel to restore the clamping height. The diaphragm must be absolutely flat when the pressure plate is fully bolted up.

Many replacement clutch/clutch remanufacturing companies or clutch specialists will custom make a 'paddle' clutch to suit any application.

Note that cerametalic clutch plates can be quite hard on flywheel and clutch cover friction surfaces. The

GENERAL INFORMATION

solution to this issue is to use a steel flywheel and a clutch cover which has a steel insert in it, as opposed to a cast iron one. AP offers both types of clutch cover.

LIGHTENED FLYWHEELS

The standard Jaguar flywheel is made out of cast iron and is very heavy, to say the least. Removing material from the flywheel to lighten it is a good idea, provided the cross sectional thickness is not reduced too much, thus weakening it.

Warning!/**Caution!** - A weakened flywheel can shatter and cause serious injury and/or mechanical damage. To avoid this, the flywheel first needs to be crack tested by an engine machine shop or an engine reconditioner (or get it X-rayed). It can then be turned down in a lathe. Only the back of the flywheel can be machined. An experienced turner will machine the flywheel making sure that reasonable radii, rather than sharp corners (potential failure points), are incorporated. The weight of the flywheel can be nearly halved provided the work is done carefully. Remember that removing material from the vicinity of the outer diameter of the flywheel has more effect than removing material from the inner diameter area (force times distance applies). The cross-sectional thickness of any XK flywheel must never be less than 0.400in/10.0mm from the flange location to the ring gear. Leave the flange that bolts up to the crankshaft at the standard thickness.

Without doubt, the lightest flywheel arrangement is made up using an automatic transmission flex plate. If getting the flywheel and clutch arrangement as light as possible is the aim, fitting a twin or triple plate aftermarket clutch in conjunction with a flex plate can't be beaten. The flex plate is really only there to supply a ring gear for starting purposes. Clutches can be mounted by various means, but the usual method involves machining a supplementary plate (steel) which is bolted onto the back of the crankshaft along with the flex plate. The twin or triple plate clutch assembly is bolted to this supplementary plate. This means that you end up with a flex plate and a high-tensile steel disc about 10-11 inches/255-280mm in diameter (depending on the size and style of the aftermarket clutch being fitted), about $7/16$-$1/2$in/10-12mm thick. The flex plate is bolted to the supplementary steel plate using about 10 small $1/4$in NF/6mm bolts near the periphery to make sure that the flex plate runs straight and true. This sort of arrangement is now quite common practice for racing cars.

It's not impossible to get the whole clutch, supplementary plate, and flex plate under an all up total weight of about 20lb/9kg, but it be more or less 'bullet proof', and operate with a minimum of strain/loading on the rear main thrust washer of the crankshaft. However, buying the components, and having them made up and fitted will be expensive.

TIMING CHAINS

Timing chains do wear over time and any modified Jaguar engine being built needs to have new genuine timing chains and guides fitted. Use lockwires on the bolts that hold the camshaft sprockets in place, and stuff rags down and around the sprockets to stop the bolts falling into the engine when they are undone. If the bolts do fall into the engine you'll probably have to strip the engine to retrieve them!

OIL PUMP

Caution! - When rebuilding one of these engines you should always fit a new oil pump. In spite of being immersed in oil, these components do wear because particles of dirt and swarf pass through them before being removed by the oil filter system. Essentially XK oil pumps have unfiltered oil flowing through them!

Caution! - Always pack a new oil pump with petroleum jelly so that it will start to prime immediately the engine is turned over. With an all new engine, remove the spark plugs and spin the engine on the starter until there is full oil pressure registering on the gauge. Then, and only then, refit the spark plugs and start the engine. If any XK engine is started without good oil pressure it will be damaged. These engines will not tolerate being run without good oil pressure (big end and main bearing failure will result).

Caution! - If an XK engine is left sitting for any length of time (3 months, or more), it's always a good idea to turn the engine over with the sparkplugs out just to make sure there is oil pressure present before firing it up. You just cannot afford to start one of these engines and run it with no oil pressure.

SUMP (OIL PAN)

The standard sump (oil pan) is quite acceptable for most high performance applications. There is enough oil capacity and the standard baffling is ideal.

Although these engines get fitted to all manner of vehicles, most installations are similar - in terms of the lateral and vertical planes - to original installations. However, in some cases the engine will not be angled as it originally was: wet sumped replica D-Types are an obvious example. **Caution!** - Where an engine's attitude is changed from original you must be aware that the oil will lay in the sump differently and there may be insufficient depth over the pick-up

SPEEDPRO SERIES

Take coolant from the points marked 'A' on the inlet port side of the cylinder head (from the top of the slot).

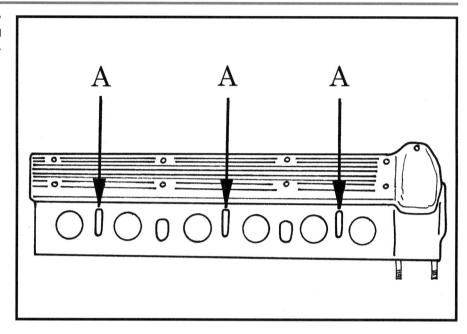

Three small bore pipes lead into one large pipe which is the same diameter as the radiator's inlet pipe. This device can be made out of mild steel tubing and fusion welded or braised. The ends of the pipes must have raised lips to prevent the hoses from coming off when the cooling system is under pressure.

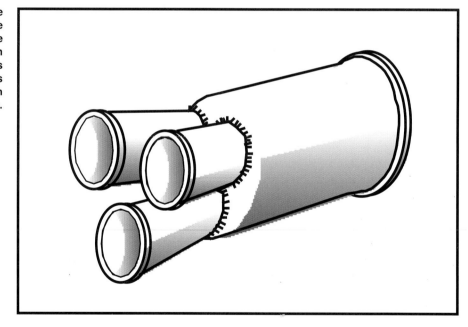

and/or other problems. In such cases the sump (aluminium or steel) can be altered to suit the new attitude of the engine.

If you're making any modifications to the sump, take care not to reduce the oil capacity from original (if anything, it should be increased). In fact, generally following the original design criteria is a sound principle. Make sure that the reservoir and oil pickup, for example, are in the original position relative to each other (i.e., the base of the sump is parallel to the ground and the oil pickup pipe is the standard distance away from it ($1/8$ to $3/16$ of an inch/3.175-4.76mm) and in the middle of the reservoir as per standard).

Jaguar made cast aluminium and pressed steel sumps for XK engines,

GENERAL INFORMATION

the latter variety are found on later engines. The advantage of the steel sump is that it is generally easier to modify than an aluminium one. Any steel sump which is modified should be Mig welded using 12 or 14 gauge mild steel sheeting (as opposed to the thinner 16 gauge sheeting that is available). Any aluminium sump that is modified will have to be Tig welded and the material thickness of the aluminium sheeting used not less than 3/16 of an inch/5mm thick. Tig welders can be a bit thin on the ground in places, so if any Jaguar sump is to be modified, obtaining a steel original and modifying that is recommended. Mig welding is also a lot cheaper than Tig welding: altering an aluminium sump can become quite expensive.

Whenever either type of sump is to be altered the sump must be bolted onto block while it's being welded to prevent distortion.

OIL COOLER

While they're often fitted, oil coolers are rarely necessary, unless the engine is to be used for endurance racing or the local ambient temperature is particularly high. If the oil temperature climbs too high without the engine being subjected to full throttle and high load work, there is no alternative but to fit an oil cooler. Try the engine without one first, though, and monitor the temperature closely to see what the trend is. Every application is different.

Caution! - Fitting an oil temperature gauge to an XK Jaguar engine being used in competition is essential. Running an XK Jaguar engine with the oil temperature near the limit is asking for trouble: if it does get too high and it is not noticed, the engine will be ruined.

Caution! - Keep the oil level at the full mark and do regular oil changes. If the engine is used for high rpm racing work, change the oil every 300 racing miles or when it's getting dirty (as can be seen on the dipstick). If the oil gets overheated, change it immediately (before the next race if necessary). XK engines are not forgiving when it comes to oil that is not in perfect condition. Many road going cars end up with a big end bearing knock if the oil is not changed at regular intervals, even if the engine is turning no more than about 3000rpm.

Caution! - These are long stroke engines with relatively heavy internal components. Having good bearings, journals running the correct clearances and clean, high quality oil between the bearings and journals is vital. These engines must be able to maintain the minimum hot oil pressure rating of 40lbs per square inch. Most engines will do this if the engine bearing clearances are not excessive and the oil pump is a brand new one.

Caution! - Never rev one of these engines until the oil temperature is up to normal running temperature. This almost always means running the engine for at least five minutes at a low speed and under low loading. Any amount of high rpm running resulting in a lack of oil between the bearing shells and the journals will result in engine bearing damage; damage which can only be repaired by replacing parts. Take no risks with these engines and they will reward your diligence with reliability second to none. Abuse one of these engines and it will fail very quickly.

COOLING SYSTEM

In competition use, XK engines have had their fair share of cooling system problems over the years. **Caution!** - For any modified XK engine, careful monitoring of the water temperature gauge is essential until the capability of the cooling system under all operating conditions is proven. Cooling problems frequently occur on engines which have non-standard inlet manifolds. Fitting a very good quality water temperature gauge is highly recommended (as is looking at it often!).

Caution! - Overheating can have serious implications for these engines. Once an XK engine has been grossly overheated the piston rings will lose their tension and the engine power will fade. The only solution is to replace the complete set of rings. This problem may not be easy to diagnose, however, because the engine won't necessarily burn oil (the oil rings are not generally affected by overheating).

There are a few engine installations that are difficult to cool after modification, such as MkIIs and S-Types, for example. These cars don't have large cross-flow radiators like later Jaguars, and there's less capacity for passing large volumes of air through the radiator. There isn't much you can do about the grill size, of course, but there is quite a bit that you can do about what's behind it. A point to remember here is that the XK Jaguar engine develops optimum power between 70-75 degrees C/158-170 degrees F coolant temperature and, if you want to keep consistent lap times, you need to keep the coolant temperature between 65-75 degrees C/150-170 degrees F. Keeping the coolant temperature to 65 degrees C/150 degrees F is very acceptable if you can do it.

Fitting a shrouded, large-bladed, viscous coupling engine-driven fan would be acceptable, as would a large capacity, thermostatically or manually-operated electric fan fitted in front of the radiator. An engine-driven fan needs to be as close to the radiator core as possible without actually touching it. Consider 1/2in/12mm to

SPEEDPRO SERIES

A custom-made cross-flow aluminium radiator can be made with a baffle fitted into one side tank (A). This baffle directs the coolant flow back across the radiator in front of the air stream twice.

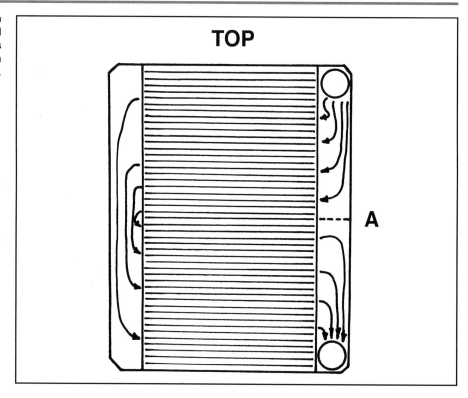

be the minimum and 1in/25mm the maximum. If the fan is manually or thermostatically-controlled, it needs to come on at 75 degrees C/170 degrees F to avoid a further increase in coolant temperature. Irrespective of anything else, reduce your speed the instant the coolant temperature reaches 90 degrees C/194 degrees F, but don't stop unless coolant loss is occurring. The temperature is much more likely to reduce quickly and safely (for the engine) with the water being circulated but without the heat generated by the use of full throttle operation.

There are some instances, however, where fitting a new fan to a modified engine results in a cooling system that is still marginal at best, and cooling problems can still be experienced.

Even with an engine block/cylinder head that has been chemically cleaned, and a new standard radiator fitted, cooling problems can still be experienced.

One expensive solution is to have an all aluminium cross-flow radiator custom made to fit the installation. The core needs to be at least the same square area as the original vertical flowing one, though it could be a bit longer in the vertical plane as there wouldn't be a top or bottom tank. There isn't a lot of extra room in the original installation, of course, and the radiator can only be so large if it's to fit into the standard original location.

Using a cross-flowing radiator

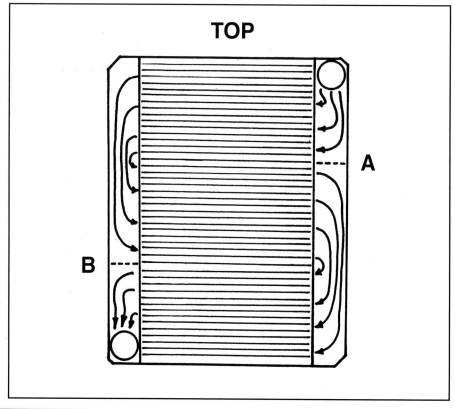

A custom-made cross-flow aluminium radiator can also be made with a baffle fitted in each side tank (A and B). The baffles direct the coolant flow back across the core three times.

GENERAL INFORMATION

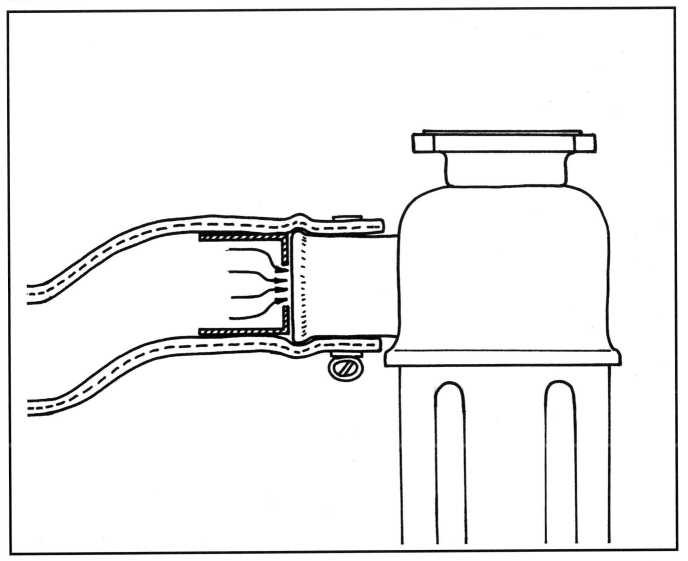

Solid restrictor shown fitted in the hose adjacent to the radiator inlet. If the engine overheats bore out the hole in the restrictor to a larger size.

allows you to do something that isn't possible on a vertical flow one, and that is to pass the coolant in front of the air stream two or three times within the same sized core. Such a radiator needs to be made out of $2^1/_4$in - $2^1/_2$in thick aluminium coring, and have side tanks which have appropriate baffles to achieve two or three passes of water through the core. The three pass system is easiest because the standard radiator hoses can still be used. The two pass system requires either a custom top or bottom hose arrangement. The accompanying schematic diagram outlines the requirements.

It's vital that a car powered by a modified XK engine be fitted with a brand new radiator and a new water pump (later, metal pressing impellor water pumps are better). Maximum possible cooling efficiency is never better than when the internal passageways of the radiator (and the engine) are clean and clear, and all of the cooling fins are in place and undamaged.

With many installations, even with all new parts, the engine can still overheat. One possible solution (only feasible if you're using triple Webers/Dellortos, however) requires

107

SPEEDPRO SERIES

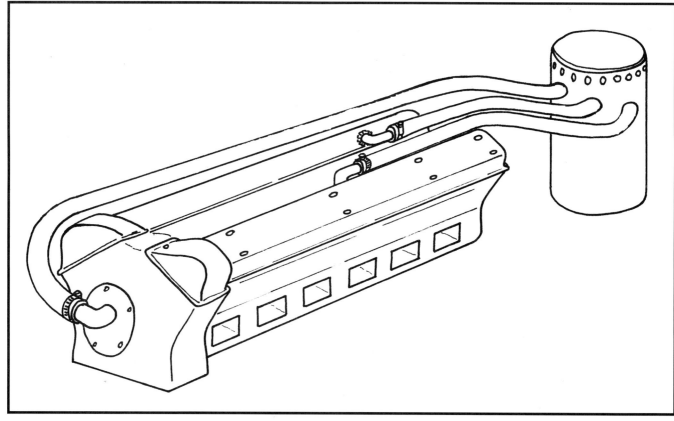

Breather pipe take offs at three points on the cylinder head/cambox covers. All three breather pipes are routed to a catch tank at the rear of the engine compartment.

taking water from three separate places along the side of the cylinder head (the three slots in the cylinder head casting between the three pairs of inlet ports). With three separate ¾in/19.05mm outside diameter water pipe outlets coming off the side of the cylinder head, three hose pipes are bought forward (as close as possible to the top of the inlet manifold) and routed into a smooth trifucated pipe joint. This system allows water to flow from the rear, middle and front of the cylinder head, into the radiator's single pipe fitting. Keep the water hoses as low as possible so that the radiator's inlet pipe fitting is the highest point of the cooling system. Hot water is not easy to pump downward!

With the three outlet system there is no provision for a thermostat and this is problematical for road cars. Engines used solely for racing don't need a thermostat because the coolant will get up to temperature quickly and, because the engine is not being started and run every day, wear caused by cold running is not really an issue.

For road going cars without a thermostat, the solution to the engine running too cold is to put a restrictor plate in the top hose that goes directly to the radiator. Start off with a 1in/25.4mm diameter hole in the plate and work up in ⅛in/3mm diameter increments until the engine temperature stablises at a reasonable running temperature (167 degrees F/75C, is fine). For competition use of the same engine, simply remove the restrictor plate at the venue, race the car and then replace the restrictor plate before leaving.

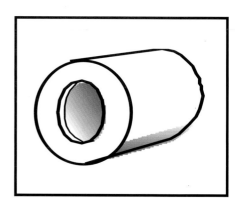

The restrictor fits in the radiator hose adjacent to the radiator inlet pipe. The restrictor plate can be a short length of tubing with an end cap with a hole in it.

GENERAL INFORMATION

The restrictor plate is a short piece of tubing of around the same diameter as the inside of the top radiator hose about 1.5in/38.1mm long and with a plate welded onto it. The short piece of tubing must fit closely into the top radiator hose, but not so tightly that it's difficult to remove.

Alternatively, a restrictor to fit the inside diameter of the top radiator hose can be turned from a solid bar of aluminium. The hole through the restrictor can be progressively opened out from 1in/25.4mm as previously described.

CRANKCASE VENTILATION

With a reasonable piston to bore clearance these engines don't normally have any crankcase ventilation problems.

The standard set up, i.e., a tube from the breather housing at the front of the engine, routed upwards along the side of the camshaft cover (pipe inclines as it goes away from the breather housing) and into a catch tank at the rear of the engine, is usually sufficient. This effectively means that the breather pipe vents to atmosphere. Check the catch tank for oil after each race. If this doesn't provide enough crankcase ventilation, weld a short length of 3/4 inch (19mm) diameter aluminium pipe into each camshaft cover adjacent to the centre of the cylinder head (the spark plug side of each cam cover). From these stubs, feed two hoses back into the catch tank. The position of the pipes needs to be near the centre of the cam covers, or slightly towards the back of the engine. Take the hose from the breather housing along the inlet side of the engine, rather than the exhaust side, unless the hose can be routed under the exhaust pipes and shielded from the heat by a piece of aluminium plate.

ENGINE TUNE-UP

Correct tuning is a vital feature of building a successful engine. Few engines go well if they have not had a substantial amount of time and effort devoted to making sure that all of the various aspects of the engine are working as they should. Get just one aspect of an engine wrong, and often the whole thing is wrong. Tuning an engine can be very time consuming, but there's no getting out of it. It has to be approached with a 'no stone unturned' policy because, for the sake of a simple setting, 10, 20, 30 or even more bhp will be lost.

The settings detailed throughout this book are all settings to which XK engines are known to respond. Although these settings should be used initially, they must be thoroughly checked out before the engine is subjected to sustained high revolution, fully loaded work. It only takes having too much compression for the octane rating of the fuel being used, and too much spark advance, to damage one of these engines. Since XK engines are not light, removing them from the car for stripdown and repair is not something you want to do very often!

The foregoing is not to say that all settings need to be conservative. Running less ignition advance than suggested to preclude pinking (pinging), or setting the air/fuel mixture extra rich just to make sure that the engine never runs lean, will mean that you lose power. Tuning an XK engine involves making sure that **everything** is **exactly right**, not just nearly right. To this end, scientific analysis is required, as is the use of a rolling road. Why guess the mixture strength and amount of ignition advance for example when you can have the engine checked and set correctly. With the basic setting built into the engine, and the engine run for a suitable amount of time so that everything has settled in, a trip to a rolling road should allow all of the settings to be narrowed down to exactly what is required. Rolling road sessions can be very informative because the engine is being monitored for mixture strength all of the time.

With the car on the rolling road and the mixture checking/scientific diagnostics connected to the exhaust pipes, the idle mixture can be checked. Aim for a reading of 0.95 Lambda/14.0 to 1 air-fuel ratio and 1.6% CO, to 0.88 Lambda/13.0 to 1 air-fuel ratio and 3.8% CO, and consider these last figures to be getting quite rich and the definite maximum. If you're tuning a road going car, you want to be looking at the leanest setting possible, such as 0.95 Lambda/14.0 to 1 air-fuel ratio. 1.6% CO, that will have the engine idling smoothly and consistently.

Carburettor spindle/butterfly synchronisation, and an identical amount of fuel in the fuel bowls are absolutely vital. Don't just set the float levels in the physical sense, check the amount of fuel in each bowl.

The idle mixture strength is set with the engine idling and not under load. Note that with SUs, the optimum settings is a fine point (which is fairly easy to find) and the engine will run lean if you wind the main jets up too much, or rich if you wind them down too far.

It also pays to measure the distance from the 'bridge' of the carburettor to the tops of the main jet in each carburettor with the tail of a vernier caliper, and set each distance to within 0.002in/0.05mm of each other. Each time an adjustment is made, either richer or leaner, check that the distances are identical on all carburettors. It does mean frequently taking off the piston and suction chambers, but it's probably the best way to make sure that the carburettors are flowing equal amounts of fuel.

The springs in the dashpots must also be of the same tension, both at the installed height and the fully compressed height (maximum vertical piston movement). This is something which is rarely checked, but it can make quite a difference to how well an SU-equipped engine runs. The only way to check them is to test the force exerted by all of the springs at the same heights (check HD8s at 3.875in and 2.000in and HD6s at 2.635in and 1.250in). They must all be equal at both listed heights. If one spring is weak it will allow that piston to 'ride' higher than the others for the same given airflow through the carburettor. This, along with the fuel level in the carburettor bowls, the position of the needles in the pistons, the distance the main jet is down from the bridge and the shape/sizes of the needles, is what causes the differences in mixture strengths between cylinders/cylinder groups.

The main jets and the needles must be as new if there is to be any hope of getting the engine perfectly tuned. Abnormal wear in these components can result in slightly different mixture strengths between carburettors, in spite of the adjustment procedure being carried out correctly. If these components all wore in the same places and by the same amounts, it wouldn't matter so much, but they don't. Contact Burlen Fuel Systems in the UK for details of the availability of all SU parts.

Although it isn't impossible to get an XK Jaguar engine to idle with an air/fuel mixture ratio of 0.94 Lambda/13.8 to 1 air-fuel ratio/2.0% to 0.92 Lambda/13.5 to 1 air-fuel ratio/1.9% CO to 0.92 Lambda/13.5 to 1 air-fuel ratio/2.6% CO on SUs, Webers or Dellortos, this amount of leanness may cause problems (such as hesitation when the throttle is opened) for SUs. This hesitation may be due to the initial mixture being too lean, but it's more likely due to the needles, and altering these may be the only way of getting this aspect of tuning correct.

To tune an engine for maximum power it needs to be run on the rolling road dyno with a mixture strength of at least 0.85-083 Lambda/5.0-5.9% CO and 38 degrees of total ignition advance before a maximum power reading is taken. With 0.85 Lambda/5.0% CO as a starting point, progressively make the mixture richer until the power reading starts to drop away, then go back a step to the last maximum power reading. For example, you may find that your engine produces maximum power at 0.84 Lambda/5.4% CO. To find this setting means that you will have taken the mixture to 0.83 Lambda/5.9% CO and registered a power reduction (1-4 bhp). The instant the power reduces, you'll know you've gone too far (too rich). Going back a step to the previous setting 0.84 Lambda/5.4% CO will give optimum power with best possible fuel use.

To run a Jaguar engine at high rpm (up to but not exceeding 6000rpm) you need to set the air/fuel mixture on the rich side of the optimum power range, not the lean side (less than 0.85 Lambda/5.0% CO). This is to keep the internal engine componentry as cool as is possible while not missing out on any power and efficiency.

With the ideal mixture strength found, the next step is to increase the ignition a degree at a time from the 38 degrees. Aim to use the least amount of advance that develops maximum power, and consider 44 degrees the absolute limit for a Jaguar engine. This approach is designed to reduce the risk

Lambda	Air/Fuel	% CO
0.80	11.8	8.0
0.81	11.9	7.3
0.82	12.0	6.5
0.83	12.2	5.9
0.84	12.4	5.4
0.85	12.5	5.0
0.86	12.6	4.85
0.87	12.8	4.35
0.88	13.0	3.8
0.90	13.2	3.3
0.91	13.4	2.85
0.92	13.5	2.6
0.93	13.7	2.15
0.94	13.8	1.9
0.95	14.0	1.6
0.96	14.1	1.4
0.97	14.3	1.0
0.98	14.4	0.8
0.99	14.6	0.6
1.00	14.7	0.5
1.01	14.8	0.6
1.02	15.0	0.3
1.03	15.1	0.15
1.04	15.2	0.2
1.05	15.4	0.15

Lambda, air/fuel, CO conversion table.

of engine damage due to too much ignition advance with a mixture which is too lean. The ignition is advanced only after the mixture strength has been finalised.

Expect a well-tuned Jaguar engine to give a cruise CO reading in the vicinity of 0.95 - 0.92 Lambda/1.6-2.6% CO. Although high rpm is possible with quite a lean mixture, it's best not to have it set too lean if sustained high rpm (4500rpm) is being used continuously with partial throttle opening (piston crown failure is a possibility).

In the interests of reliability, keep the water temperature down, as recommended earlier in the book, with 65-75 degrees C/150-175 degrees F being ideal.

GENERAL INFORMATION

A road going engine should be set to the low side of the percentage CO recommendations in the interests of fuel economy but that mixture might not be all that suitable for sustained high rpm high load cruising.

When testing an engine on a rolling road keep an eye on the water temperature at all times. It's very easy to overlook this when dyno testing and the temperatures can get out of hand very quickly. All rolling road establishments use large portable blowers to keep a large volume of air moving through the radiator, but they are never usually good enough to keep the engine cool, especially a modified engine that generates more heat than standard. Once the water temperature gets to 80 degrees C/175 degrees F, stop testing.

Because there are three mixture measuring systems used (air/fuel ratios, CO (Carbon Monoxide) and Lambda), I have included the accompanying conversion table to assist with cross-referencing.

LUCAS FUEL INJECTION EQUIPPED ENGINES

While not many engines use this equipment, it's useful to bear in mind that fuel injection systems, like the Lucas L-Jetronic, can have the air/fuel mixture altered to quite a degree for not a lot of money. Matched injectors are required, however, as there is just too much of variation in the spray patterns and the amounts of fuel that the standard injectors deliver. There can be up to 35% difference in the amount of fuel delivered, injector to injector. This means that there could be up 35% variation in the amount of fuel being delivered to each cylinder versus equal amounts of air. This is a worst-case scenario, of course, but it's not all that uncommon. In the UK matched injectors are available from Mech Motorsport (Tel: 01242 243385, www.rollingroad.co.uk) or Power Engineering (Tel: 01895 255699, www.powerengineering.co.uk).

Fitting a rising rate fuel pressure regulator is also recommended. The FSE regulator, as made by Malpassi, is ideal for XK engines. The standard air flow meter (richness/leanness) can be adjusted by increasing or decreasing the spring tension by moving the ratchet in the adjustment mechanism. There is a limit to this, however, because the richness will be increased uniformly across the rev range. This means that while it would be quite possible to set the mixture strength for optimum top end power and performance, at idle the mixture will probably be grossly rich.

The correct way to obtain an ideal fuel curve with an L-Jetronic EFI is to have the ECU altered, which is neither easy nor cheap. For sound advice on these ECUs contact Mark Adams of Tornado Engine Management Systems on (UK phone number) 01694 720144, or by e-mail at mark.adams@bjds.com.

2.4 AND 2.8 LITRE XK ENGINES

While not the subject of this book, these engines are very similar to the larger versions of the family. They weren't very successful, however, because they weren't powerful enough for the weight of the cars, and fuel consumption was poor. Both engines respond to the sort of tuning criteria outlined in this book. The crankshafts and connecting rods are pretty 'bulletproof' and, if forged pistons are fitted, both will stand 7500rpm. The 2.8 litre engines, for example, were available with high dome 9.0:1 compression pistons and, with block and cylinder head planing, it's quite possible to get the compression up to 9.5:1-10:1. The 2.8 litre engine in particular is the more powerful of the two.

The 2.4 and 2.8 litre engines had the same cylinder heads as the 3.4, 3.8 and 4.2 litre engines, and the modifications listed in this book apply.

Jaguar never attempted to increase the bore of the 2.8 litre engine (made it a 3.4, for example) in the same way that it did for the 4.2 litre engine, though I'm sure such an engine would have been excellent. A 2.8 litre engine bored out by 0.060in/1.5mm becomes 2.9 litres. Some of these engines were sleeved, however, and it's not advisable to go this far with sleeves (0.040in/1.0mm oversize is maximum). Those blocks which weren't sleeved can go out to 0.060in/1.5mm or more (up to 0.080in/2.0mm with custom-sized pistons). It's a good idea to have the block ultrasonically tested first just to make sure that there are no core shifts in it. Everything else, such as camshafts, carburettors, ignition, exhaust systems, and all of the other associated things that apply to the 3.4, 3.8 and 4.2 litre engines can apply to the 2.8.

The 2.4 litre engine is a little different in that the cylinder head doesn't need to be ported out to the same extent as those on a 3.4, 3.8 or 4.2 litre engine. Three 40mm DCOE or 40mm DHLA Dellorto carburettors are often all that are required on these engines, and the exhaust pipe diameter doesn't need to be as large. 1⁵⁄₈in for the primary pipes and 1³⁄₄in for the two main pipes will usually be adequate.

NOTES

NOTES

Also from Veloce Publishing -

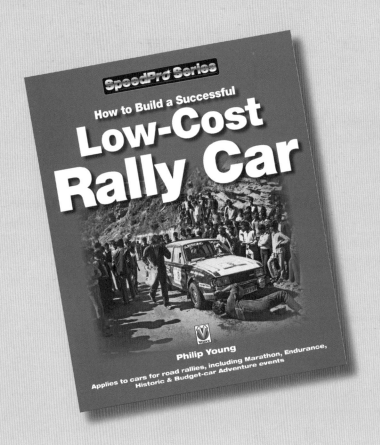

How to Build a Successful Low-Cost Rally Car For Marathon, Endurance, Historic and Budget-car Adventure Road Rallies
Phillip Young

Simple, cost-effective, basic and reliable tips to ensure that any rally car stands a chance of reaching the finishing line. If you are planning a road-based rally, don't even think of leaving home before reading this book and implementing the tried and tested mods it describes so well.

V4208 • Paperback • 25x20.7cm • £16.99
• 96 pages • 150 colour pictures
• ISBN: 978-1-845842-08-6
• UPC: 6-36847-04208-0

How To Build & Power Tune Weber & Dellorto DCOE, DCO/SP & DHLA Carburettors 3rd Edition
Des Hammill

Packed with information on stripping and rebuilding, tuning, jetting, and choke sizes. Application formulae help you calculate exactly the right set-up for your car. Covers all Weber DCOE & Dellorto DHLA & DCO/SP carburettors.

V275 • Paperback • 25x20.7cm • £19.99
• 128 pages • 180+ mainly colour pictures
• ISBN: 978-1-903706-75-6
• UPC: 6-36847-00275-6

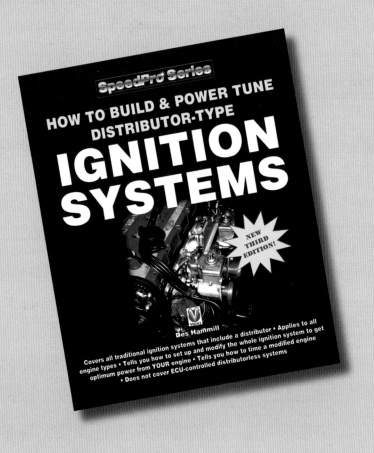

How to Build & Power Tune Distributor-type Ignition Systems
Des Hammill

How to build an excellent ignition system and optimise the ignition timing of any high-performance engine. Applies to four-stroke engines with distributor-type ignition systems (including electronic ignition modules). Does not cover engines controlled by ECUs.

V4186 • Paperback • 25x20.7cm • £16.99
• 80 pages • 80 colour photos & illustrations
• ISBN: 978-1-84584-186-7
• UPC: 6-36847-04186-1

The Competition Car Data Logging Manual
Graham Templeman

At last! A practical handbook on how to choose and operate data logging equipment and get the full benefit from what it tells you. Aimed at the amateur competitor, the book covers hardware and software, and takes over where the manufacturers' instructions leave off.

V4162 • Paperback • 25x20.7cm • £19.99
• 128 pages • 120 colour pictures
• ISBN: 978-1-845841-62-1
• UPC: 6-36847-04162-5

Index

Block threads 16, 17
Block thrust washer register 21
Bores 10-12, 14, 15

Camshafts, high performance 58, 59
Camshaft timing 64-68
Camshaft timing marks 64, 65, 68
Camshaft types (durations and phasing) 56-59, 92
Centrifugal advance 70
Circlips 12, 14
Clutch driven plates 94
Clutch pressure plates 93, 94
Compression ratios 15, 16
Connecting rod 'bearing crush' 25, 26
Connecting rod big ends 23-27
Connecting rod big end sizes 25, 26
Connecting rod bolts 23, 25
Connecting rod little ends 23, 25
Connecting rods 22, 23
Connecting rod straightness testing 22, 23
Contact breakers 69
Crack testing 22, 94
Crankcase ventilation 98, 99
Crankshaft journals and sizes 27, 28

Crankshafts 22, 26-28
Crankshaft thrust washers 28, 29, 93
Cylinder bores 10-12, 14, 15
Cylinder blocks 10-12, 14, 15
Cylinder head gaskets 11, 14-16
Cylinder head refacing 51-53
Cylinder head studs 16, 17
Cylinder head types 30, 31

Dellorto carburettors 85-87
Dellorto jetting 85, 86
Distributor springs 70, 71

Electronic distributors 69
Engine balance 92, 93
Engine cooling system 97-99
Engine tune up 99, 100
Exhaust pipe configurations 88
Exhaust ports 31-34
Exhaust systems 88-91
Exhaust valve opening points 58
Exhaust valve seat inserts 32-34
Exhaust valve seats 33, 34

Flywheels 94, 95
Freeze plugs 10, 17, 18

Gudgeon pins 12, 14

Ignition systems 69-77
Ignition timing marks 71-74
Ignition timing setting and testing 75-77
Ignition timing total advance 76, 77
Inlet ports 34-47
Inlet valve closing points 58
Inlet valves 49-51
Inlet valve seats 48, 51

Main bearing tunnels 18-21
Main bearing tunnel sizes 18-21
Main 'bearing crush' 18-21
Main caps 10
Main pipe diameters (exhaust systems) 90

Oil coolers 95-97
Oil pans 95
Oil pumps 95

Piston circlips 12, 14,
Pistons 10-12, 14-17
Piston to bore clearances 11

Piston ring gaps 14
Piston ring grooves 12-14
Piston rings 12-15
Piston wear 11, 12
Pre-ignition 70, 76, 77
Primary pipes (exhaust systems) 88, 90

Silencers (exhaust systems) 90, 91
Sleeved or linered blocks 16, 17
Spun bearings 18
Static advance 70, 75
SU carburettors 78-83
SU carburettors (modified) 82, 83
SU carburettors (twin 1 3/4 inch) 79
SU carburettors (twin 2 inch) 79, 80
SU carburettors (twin HIF 7s) 81
SU carburettors (triple 2 inch) 81
Sumps 95

Tappet guides 53
Teflon buttons 14
Timing chains 95
Total ignition advance 76, 77

Vacuum advance 69-71, 75
Valve clearances 92
Valve collets/keepers 60
Valve guide inserts (K-Line) 35
Valve guides 34-37
Valve overlap 58
Valves 31, 49-51
Valve springs 60-63
Valve springs (checking poundage) 62, 63
Valve springs (stronger than standard) 61, 62
Valve stem seals 63, 64

Weber carburettors 85-87
Weber jetting 86